ORGANOMETALLIC POLYMERS

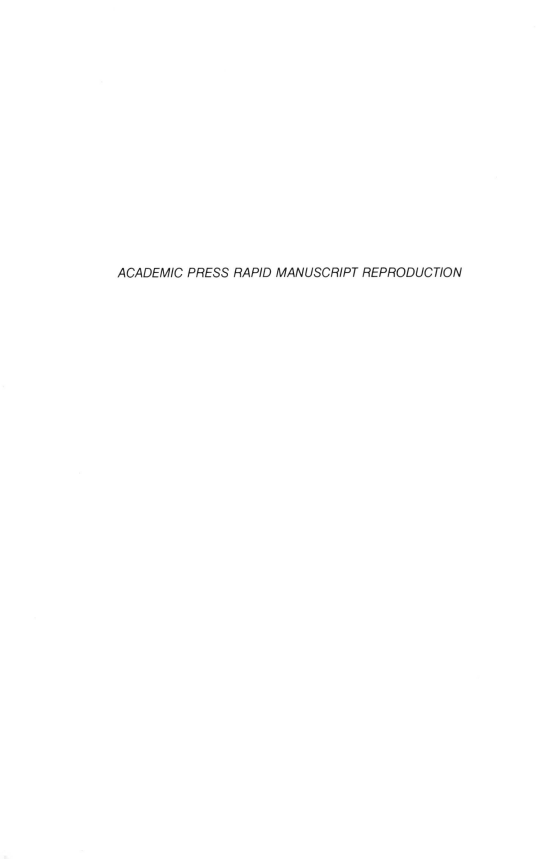

ORGANOMETALLIC POLYMERS

Edited by

Charles E. Carraher, Jr.
Wright State University
Dayton, Ohio

John E. Sheats
Rider College
Lawrenceville, New Jersey

Charles U. Pittman, Jr.
University of Alabama
University, Alabama

Academic Press
New York   San Francisco   London   1978
A Subsidiary of Harcourt Brace Jovanovich, Publishers

ACADEMIC PRESS, INC.
111 Fifth Avenue, New York, New York 10003

United Kingdom Edition published by
ACADEMIC PRESS, INC. (LONDON) LTD.
24/28 Oval Road, London NW1 7DX

Library of Congress Cataloging in Publication Data

Symposium on Organometallic Polymers, New Orleans,
1977.
Organometallic polymers.

The symposium took place during the national
meeting of the American Chemical Society in
New Orleans, La., 1977.
1. Polymers and polymerization—Congresses.
2. Organometallic compounds—Congresses.
I. Carraher, Charles E. II. Sheats, John E.
III. Pittman, Charles U. IV. American
Chemical Society. V. Title.
QD380.S93   1977   547'.8427   77-20108
ISBN 0-12-160850-6
PRINTED IN THE UNITED STATES OF AMERICA

CONTENTS

ORGANOMETALLIC CONDENSATION POLYMERS

SYNTHESIS AND PROPERTIES OF POLYMER-BOUND
TRANSITION METAL CATALYSTS

ANTIFOULING APPLICATIONS OF TIN CONTAINING ORGANOMETALLIC POLYMERS

NEW DEVELOPMENTS IN ORGANOSILICON POLYMERS

LIST OF CONTRIBUTORS

M. Akhtar, Department of Chemistry, University of Pennsylvania, Philadelphia, Pennsylvania 19174

M. N. Alexander, US Army Materials and Mechanics Research Center, Watertown, Massachusetts 02172

Harvey R. Allcock, Department of Chemistry, The Pennsylvania State University, University Park, Pennsylvania 16802

R. Atwal, School of Chemistry, Thames Polytechnic, Woolwich, England

J. C. Bailar, Jr., Department of Chemistry, University of Illinois, Urbana, Illinois 61801

Douglas G. Borden, Research Laboratories, Eastman Kodak Company, Rochester, New York 14650

Charles E. Carraher, Jr., Department of Chemistry, Wright State University, Dayton, Ohio 45431

V. J. Castelli, David W. Taylor Naval Ship R&D Center, Annapolis, Maryland 21402

C. K. Chiang, Department of Physics, University of Pennsylvania, Philadelphia, Pennsylvania 19174

Eui-won Choe, Celanese Research Company, Summit, New Jersey 07901

M. J. Cohen, Department of Physics, University of Pennsylvania, Philadelphia, Pennsylvania 19174

C. B. Cook, School of Chemistry, Thames Polytechnic, Woolwich, England

Jamie Corredor, Department of Materials Science and Engineering, Washington State University, Pullman, Washington 99164

B. R. Currell, School of Chemistry, Thames Polytechnic, Woolwich, England

James J. Dechter, Department of Chemistry, University of California, Los Angeles, California 90024

C. R. Desper, Polymer and Chemistry Division, Army Materials and Mechanics Research Center, Watertown, Massachusetts 02172

Bruce H. Edwards, Chemistry Department, University of Massachusetts, Amherst, Massachusetts 01002

Lydia M. Frenzel, Department of Chemistry, University of New Orleans, New Orleans, Louisiana 70122

B. K. Garg, Department of Materials Science and Engineering, Washington State University, Pullman, Washington 99164

M. H. George, Department of Chemistry, Imperial College of Science and Technology, London, England

Ivan J. Goldfarb, Air Force Materials Laboratory, Wright-Pattrson Air Force Base, Ohio, 45433

Mary L. Good, Department of Chemistry, University of New Orleans, New Orleans, Louisiana 70122

S. Gourlay, US Army Medical Bioengineering Research and Development Lab, Fort Detrick, Maryland 21701

Harry B. Gray, A. A. Noyes Laboratory of Chemical Physics, California Institute of Technology, Pasadena, California 91125

Robert H. Grubbs, Department of Chemistry, Michigan State University, East Lansing, Michigan 48824

Amitava Gupta, Energy and Materials Research Section 346, Jet Propulsion Laboratory, Pasadena, California 91103

W. O. Haag, Princeton Laboratory, Mobil Research and Development Corporation, Princeton, New Jersey 08540

Gary L. Hagnauer, Polymer and Chemistry Division, Army Materials and Mechanics Research Center, Watertown, Massachusetts 02172

G. F. Hayes, Explosives Research and Development Establishment, Ministry of Defense, Waltham Abbey, Essex, England

A. J. Heeger, Department of Physics, University of Pennsylvania, Philadelphia, Pennsylvania 19174

A. Hegyeli, US Army Medical Bioengineering Research and Development Lab, Fort Detrick, Maryland 21701

James F. Hoffman, Department of Chemistry, University of New Orleans, New Orleans, Louisiana 70122

Joseph Jagur-Grodzinski, Plastics Research Department, Weizmann Institute of Science, Rehovot, Israel

A. T. Jurewicz, Mobil Chemical Corporation, Edison, New Jersey 08817

Keith C. Kappel, Department of Chemistry, University of New Orleans, New Orleans, Louisiana 70122

Frederick J. Karol, Union Carbide Corporation, Research and Development, Chemicals and Plastics, Bound Brook, New Jersey 08805

J. Kleppinger, Department of Chemistry, University of Pennsylvania, Philadelphia, Pennsylvania 19174

V. V. Korshak, Vavilov Street 28, Moscow, USSR

W. H. Lang, Princeton Laboratory, Mobil Research and Development Corporation, Princeton, New Jersey 08540

A. G. MacDiarmid, Department of Chemistry, University of Pennsylvania, Philadelphia, Pennsylvania 19174

H. G. Midgley, School of Chemistry, Thames Polytechnic, Woolwich, England

J. Milliken, Department of Chemistry, University of Pennsylvania, Philadelphia, Pennsylvania 19174

Eric A. Mintz, Department of Chemistry, University of Massachusetts, Amherst, Massachusetts 01002

Charles P. Monaghan, Department of Chemistry, University of New Orleans, New Orleans, Louisiana 70122

Mark Moran, Department of Chemistry, University of South Dakota, Vermillion, South Dakota 57069

M. J. Moran, Department of Chemistry, University of Pennsylvania, Philadelphia, Pennsylvania 19174

Eberhard W. Neuse, University of Witwatersrand, Johannesburg, South Africa

Emery Nyilas, Avco Everett Research Lab, Everett, Massachusetts 02149

Elmer J. O'Brien, Department of Chemistry, University of New Orleans, New Orleans, Louisiana 70122

Yehuda Ozari, GAF Corporation, 1361 Alps Road, Wayne, New Jersey 07470

J. R. Parsonage, School of Chemistry, Thames Polytechnic, Woolwich, England

D. L. Peebles, Department of Physics, University of Pennsylvania, Philadelphia, Pennsylvania 19174

Charles U. Pittman, Jr., Department of Chemistry, University of Alabama, University, Alabama 35486

Marvin D. Rausch, Department of Chemistry, University of Massachusetts, Amherst, Massachusetts 01002

Charlene Deremo Reese, Department of Chemistry, University of South Dakota, Vermillion, South Dakota, 57069

Alan Rembaum, Energy and Materials Research Section 346, Jet Propulsion Laboratory, Pasadena, California 91125

R. Rice, US Army Medical Bioengineering and Development Lab, Fort Detrick, Maryland 21701

L. D. Rollmann, Princeton Laboratory, Mobil Research and Development Corporation, Princeton, New Jersey 08540

Harold Rosenberg, Air Force Materials Laboratory, Wright-Patterson Air Force Base, Ohio 45433

Thane D. Rounsefell, University of Alabama, University, Alabama 35486

P. L. Sagalyn, Polymer and Chemistry Division, Army Materials and Mechanics Research Center, Watertown, Massachusetts 02172

N. S. Schneider, Polymer and Chemistry Division, Army Materials and Mechanics Research Center, Watertown, Massachusetts 02172

John E. Sheats, Department of Chemistry, Rider College, Lawrenceville, New Jersey 08648

A. Siegel, Department of Chemistry, Indiana State University, Terre Haute, Indiana 47809

Robert E. Singler, Polymer and Chemistry Division, Army Materials and Mechanics Research Center, Watertown, Massachusetts 02172

D. W. Slocum, Neckers Laboratories, Southern Illinois University, Carbondale, Illinois 62901

R. B. Somoano, Jet Propulsion Laboratory, Pasadena, California 91103

S. L. Sosin, Vavilov Street 28, Moscow, USSR

Chit Srivanavit, Department of Chemistry, University of California, Los Angeles, California 90024

Shiu-Chin H. Su, Department of Chemistry, Michigan State University, East Lansing, Michigan 48824

R. V. Subramanian, Department of Materials Science and Engineering, Washington State University, Pullman, Washington 99164

W. Volksen, Jet Propulsion Laboratory, Pasadena, California 91103

C. W. R. Wade, US Army Medical Bioengineering Research and Development Center, Fort Detrick, Maryland 21701

Robert S. Ward, Jr., Avco Medical Products, Everett, Massachusetts 02149

J. White, Polymer and Chemistry Division, Army Materials and Mechanics Center, Watertown, Massachusetts 02172

D. D. Whitehurst, Princeton Laboratory, Mobil Research and Development Corporation, Princeton, New Jersey 08540

R. Williams, California Institute of Technology, Pasadena, California 91125

T. N. Williams, Jr., Department of Chemical Engineering, Princeton University, Princeton, New Jersey 08540

W. L. Yeager, David W. Taylor Naval Ship R&D Center, Annapolis, Maryland 21402

Jeffrey I. Zink, Department of Chemistry, University of California, Los Angeles, California 90024

PREFACE

Organometallic Polymers is a volume designed to cover a wide range of topics related to organometallic polymers: their synthesis, characterization, and potential applications. For the purpose of this volume we have defined an organometallic polymer as one in which the metal-containing portion is either incorporated as an integral part of the polymer or is bound to the polymer structure by covalent bonds. Some of the materials such as the phosphonitriles and sulfur nitride polymers might be classed as inorganic rather than organometallic polymers, but are included because of their close relationship to the other types of polymers discussed.

The chapters in this volume were written by the speakers at the three-day Symposium on Organometallic Polymers, held at the National Meeting of the American Chemical Society in New Orleans, on March 20–23, 1977. The authors are recognized experts within their areas of research. Most are American but some contributions from abroad have been included. There is a mix of industrial and academic authors and a mixture of basic, theoretical, and applied research topics. The volume is divided into seven sections: vinyl polymerization of organometallic monomers, condensation polymerization of organometallic monomers, polymer-bound catalysts, applications of organotin polymers, recent developments in organosilicon polymers, phosphonitrile and sulfur nitride polymers, and coordination polymers. Each section includes one or more summary or review papers, which covers progress in the field, and includes several other papers presenting various aspects of the topic. Of the polymeric materials described only two classes, the organosilicons and the polymer-bound chromocene catalysts, have widespread commercial uses. Other materials, such as the organotin polymers, the phosphonitriles, and some of the hydroformylation catalysts, seem certain to be used widely within a few years. The others at present remain laboratory curiosities and subjects of an increasing amount of research. Potential applications include use as adhesives, antifouling paint, bacteriacides, biopolymers, catalysts, conductors, controlled-release agents, fibers, flame retardants, fungicides, haptens, implants in living tissue, paints, photo sensitizers, photostabilizers, photoresists, semiconductors, and stereospecific catalysts. Instrumental techniques for characterizing materials such as Mössbauer spectroscopy, photoacoustic spectroscopy, torsional braid analysis, thallium NMR, and X-ray crystallography are described. Considering the enormous impact the plastics and polymers formed from carbon, hydrogen, nitrogen, oxygen, silicon, and the halogens have already had on society, much more awaits the inclusion of the rest of the periodic table. In this volume polymers containing 26 other elements are described. Another symposium on the same topic is planned for 1980. We hope to have many more exciting results in the field by then.

VINYL POLYMERIZATION OF ORGANOMETALLIC
MONOMERS CONTAINING TRANSITION METALS

Charles U. Pittman, Jr.
Department of Chemistry, University of Alabama

ABSTRACT. This chapter provides a general review of the vinyl monomers, containing transition metals, which have been prepared and polymerized. The reactivity of such monomers in addition to homo- and copolymerizations is described. The Q-e scheme is used to semiempirically classify the vinyl reactivity of several organometallic monomers.

I.   INTRODUCTION.

The effect that transition metal functions exert in vinyl polymerization of vinyl organometallic monomers has just recently undergone serious study (1). Example transition metal-containing monomers are vinylferrocene *1* (2,3) vinyl-cyclopentadienyltricarbonylmanganese *2* (4), Vinylcyclopenta-dienyldicarbonylnitrosylchromium *3* (5), styrenetricarbonyl-chromium *4* (6), trans-bis(tributylphosphine) (4-styryl)pallad-iumchloride *5* (7) η^6-(2-phenylethyl acrylate)tricarbonylchrom-ium *6* (8), and η^4-(2,4-hexadiene-1-yl acrylate)tricarbonyl-iron *7* (9). The first vinyl polymerization of an organo-metallic derivative was the radical-initiated homopolymeriza-tion of vinylferrocene by

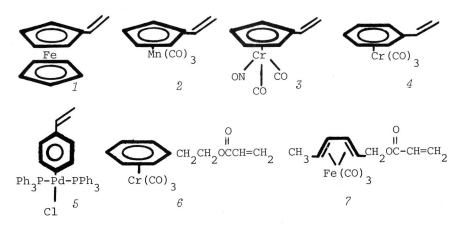

Arimoto and Haven in 1955 (10). For the next decade this area was virtually neglected in comparison with the vast attention given to organometallic condensation polymers. In this chapter, a brief review of the status of vinylorganometallic polymerizations will be given.

II. HOMOPOLYMERIZATION.

Vinylferrocene has been more thoroughly studied than any other organometallic monomer (2,3,11). Homopolymerization has been carried out using radical (2,3,11), cationic (12), and Ziegler-Natta initiators (13). Peroxides oxidize ferrocene. Using AIBN, solution polymerizations result in low molecular weights, but bulk polymerizations give higher molecular weights. Unlike most vinyl monomers, the molecular weight does not increase with a reduction in initiator concentration, but it does increase with an increase in monomer concentration (2). This anomalous behavior was explained by showing that vinylferrocene has a high chain transfer constant (C_m = 8 x 10^{-3} versus 6 x 10^{-5} for styrene at 60°) (3). Furthermore, the kinetics of homopolymerization in benzene follow r_p = k[VF][AIBN]. This rate law requires an intramolecular termination process as shown below:

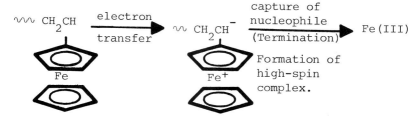

Apparently, intramolecular electron transfer from iron to the radical end occurs giving an Fe(III) end group. This behavior was subsequently supported by Mössbauer spectroscopy (11) which showed the presence of an absorption at 0.14 mm sec^{-1} that was not due to either ferrocene or ferrocenium groups. In dioxane, the usual half order dependence in monomer (bimolecular termination) was observed (11).

A high chain transfer constant to polymer in polyvinylferrocene polymerizations leads to chain branching. As the molecular weight increases, the resulting polyvinylferrocene becomes increasingly branched. Thus, vinylferrocene exhibits an unusual homopolymerization behavior which may be attributed to the influence of the organometallic function. This raises the question: How will other organometallic groups influence vinyl polymerizations?

Unusual kinetic behavior has been found in homopolymerization of *2* (14). For example, in benzonitrile $r_p = k[(2)]^{3/2} [AIBN]^{1/2}$. The explanation for this result is not yet known and kinetic studies in other solvents are in progress. Acrylic monomers of ferrocene, where the ferrocenyl moiety is insulated from the reactive vinyl group, appear to follow the normal terminal model mechanism. For example, ferrocenylmethyl acrylate, *8*, and ferrocenylmethyl methacrylate, *9*, both exhibit a half order dependence on initiator (15) (i.e., $r_p = K[monomer]^1 [AIBN]^{\frac{1}{2}}$. On the other hand, styrenetricarbonylchromium, *4*, will not homopolymerize at all, although it readily copolymerizes (6). The reason for this is unclear. Steric arguments appear invalid because *4* readily copolymerizes with *2* (6), and the very bulky 3-vinylbisfulvalenediiron, *10*, has been observed to homopolymerize (16). Monomers *5-7* each readily homopolymerize using azo initiators but no kinetic studies are

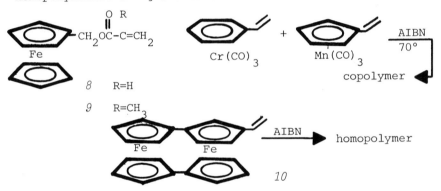

currently available. Because transition metals can far more readily undergo ionization than carbon, the potential for electron transfer mechanisms to complicate polymerization mechanisms is high.

Monomer *11* has been prepared but it would not undergo radical-initiated homopolymerization (9). Indeed, it would not copolymerize and it inhibited the polymerization of styrene and methyl acrylate. Presumably, the radical, resulting from chain attack at its vinyl group, is stable and does not permit chain propagation. Titanium allyl and methacrylate monomers *12* and *13* give only very low molecular weight materials using benzoyl peroxide initiation (17). On copolymerization with styrene only small amounts of *12* and *13* are

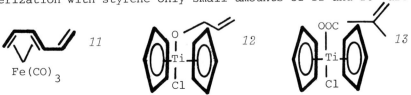

incorporated in accord with a low reactivity and high chain transfer activity for these monomers (18).

A surprising effort has been expended to polymerize ethynylferrocene *14*. Free radical, cationic, Ziegler, and anionic initiation have been tried but in most studies the resulting polymers were poorly characterized (19-25). Benzoyl peroxide initiation gives very short chains (23). Using AIBN initiation at 190°, poly(ethynylferrocene) was obtained without evidence of aliphatic C-H absorptions in the ir (24). The highly purified polymer is an insulator (σ = 4 x 10^{-14} ohm^{-1} cm^{-1}) but mixed-valence polymers were prepared by partial oxidation with agents such as iodine and DDQ and these polysalts were semiconducting (26).

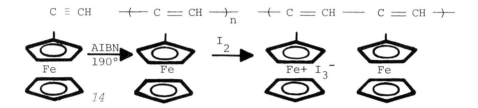

III. COPOLYMERIZATIONS.

How do organotransition metal functions effect the reactivity of vinyl groups in cationic, radical, and anionic copolymerizations? The standard way to classify the vinyl copolymerization reactivity of organic monomers has been to obtain reactivity ratios and, using these, derive the values of Q and e of that monomer. Thus, the semiempirical Q-e scheme has been employed to classify several organometallic monomers according to the electron richness (Price's polarity term e) of the vinyl group in copolymerization reactions. Using styrene as the standard comonomer (M_2), relative reactivity ratios (r_1 and r_2) have been determined using the nonlinear least squares fitting of the integral form of the copolymer equation advocated by Tidwell and Mortimer (27). The values of Q and e for organometallic monomers were then computed from:

$$r_1 = (Q_1/Q_2)\exp{-e_1(e_1-e_2)}$$
$$r_2 = (Q_2/Q_1)\exp{-e_2(e_2-e_1)}$$
$$r_1r_2 = \exp{-(e_1-e_2)^2}$$

Since a ferrocene moiety is an enormously stabilizing function toward adjacent positive charge, we thought the vinyl group should appear extraordinarily electron rich when under attack by radicals which are electron deficient. This turned out to be correct. An e_1 value of -2.1 was found for vinyl-ferrocene in styrene copolymerizations (28). How would this trend be effected if a tricarbonylmanganese group is complexed to the cyclopentadienyl ring? Will the carbonyl groups exert a strong electron withdrawing group which is felt at the vinyl group? The answer is no (28).

The values of e_1 determined for organometallic monomers *1*, *2*, and *3* in styrene copolymerizations are all large negative values, emphasizing the fact that their vinyl groups are all exceedingly electron rich (28). The cyclopentadienyl ring seems to be a very strongly e^--donating group in all cases. The para palladium function in *5* also appears to be an exceptionally strong e^--donator (7). In fact, the value of e_1

Values of e_1 for Organometallic and Standard Monomers

Fe	Mn(CO)$_3$	(MeO- ⟩-)$_2$C=CH$_2$	Cr ON|CO CO
e -2.1	-1.99	-1.96	-1.9

Ph$_3$P-Pd-PPh$_3$ Cl	N(Me)$_2$	CH$_2$=CHCN	O⟩O
e -1.62	-1.37	-.80 +1.21	+2.25

Me$_3$CCH=CH$_2$	Me$_3$SiCH=CH$_2$	Me$_3$GeCH=CH$_2$	Me$_3$SnCH=CH$_2$
e -.63	-.14	+.43	+.96

---

for *5* is -1.62, which is significantly larger negatively than that for p-dimethylaminostyrene (e = -1.37). For the main group series of organometallic monomers, R-CH=CH$_2$ (where R = Me$_3$C, Me$_3$Si-, Me$_3$Ge-, and Me$_3$Sn-) the order of e values is puzzling. As one goes from carbon through tin the vinyl groups become less electron rich (29). It is strange that the larger and less electronegative tin atom does not appear, by the Q-e scheme, to readily release electron density in radical copolymerizations. In this series the Q values were low in every case indicating no great resonance stabilization was involved.

The systematic copolymerization and Q-e classification of vinylferrocene *1* and vinylcyclopentadienyltricarbonylmanganese *2* has now been carried out (28). A range of comonomers was chosen to span a wide range of e_2 values from very electron rich to very electron deficient. Reactivity ratios were derived from a large number of copolymerizations (in most cases) and the values of e_1 for *1* and *2* were calculated. These values are listed in Table 1. A striking feature is that the Q-e scheme appears to break down when electron deficient comonomers are employed (i.e., acrylonitrile etc.). The break occurs after vinyl acetate. This phenomenon, similar to a sudden change of slope in a Hammett σ-ρ plot, signifies a change in mechanism (30). Since the Q-e scheme is defined in terms of the terminal model, it is no longer appropriate to apply the scheme if this model mechanism is not operating.

A logical interpretation for the failure of the Q-e scheme to hold, when using electron deficient comonomers, invokes competition from a charge-transfer-complex model. Since monomers *1* and *2* are extremely electron rich, the formation of charge-transfer complexes with acrylonitrile, diethylfumarate, fumaronitrile or maleic anhydride is suggested. In some cases spectral evidence confirms the existence of complexes. If the charge-transfer complexes enter the propagation sequence, then the Q-e scheme would no longer have any meaning because the Q-e scheme derivation assumes the terminal copolymerization model is operating.

Acrylic monomers of ferrocene, dienetricarbonyliron, and (η^6-phenyl)tricarbonylchromium have been synthesized and co-polymerized with common organic monomers (styrene, methyl acrylate, methyl methacrylate, acrylonitrile, and vinyl acetate). These acrylic organometallic monomers include (*6-9*) and (*14-17*). The large number of copolymerizations performed permitted the determination of the reactivity ratios for a large number of cases. These are summarized in Table 2 with references.

TABLE 1

Classification of Vinylferrocene, 1, and Vinylcyclopentadienyl-tricarbonylmanganese, 2, According to the Q-e Scheme

Comonomer M_2	e_2	Values of e_1 from Copolymerizations	
		Vinylferrocene	Vinylcyclopenta-dienyltricarbonyl-manganese
Vinylferrocene	-2.1	-	-2.6
Vinylcyclopenta-dienyltricarbon-ylmanganese	-1.99	-2.6	
N-Vinylcarbazole	-1.40	-2.4	-
p-N,N-Dimethyl-aminostyrene	-1.37	-2.2	-1.9
1,3-Butadiene	-1.05	-1.80[a]	-
	-1.40	-2.1	
N-Vinyl-2-Pyrrolidone	-0.90	-2.1	-2.9
Styrene	-0.80	-2.1	-1.99
Vinyl Acetate	-0.22		-1.62
Methyl Meth-acrylate	+0.40	-0.20[b]	
Methyl Acrylate	+0.58	-0.32[b]	-0.95
Acrylonitrile	+1.20	-0.81[b,c]	-0.95
Dimethylfumarate	+1.25	alternating	
Fumaronitrile	+1.96	alternating	alternating
Maleic Anhydride	+2.25	-0.1[d]	

[a] A value of -1.05 was reported in reference (33), but the value of -1.80 is calculated from the same experimental points using the nonlinear least squares method and the integrated form of the copolymer equation.

[b] Slightly different e_1 values are listed in the references. Values in this table were computed from data in those references as per footnote a.

[c] $r_1 \cdot r_2 = 0.024$, high alternation tendency.

[d] $r_1 \cdot r_2 = 0.003$, close to alternating.

IV.   CATIONIC AND ANIONIC INITIATION.

Using high vacuum-break seal techniques, we have been unable to obtain polymers *1*, *2*, or *3* anionically. Initiator systems included KNaph, NaNapth, BuLi, LiAlH$_4$, PhMgBr, and EtMgBr. Since vinyl monomers are so electron rich (i.e., Q-e) scheme), this behavior is not surprising. There is some evidence that polyvinylferrocene is a product of the reaction of vinylferrocene with excess soidum in HMPA (22). The successful anionic homo- and copolymerization of ferrocenylmethyl acrylate (14) and methacrylate (15) has been accomplished (23). Using LiAlH$_4$ initiation, high molecular weights (to 7.5 x 10^5) were achieved at narrow distributions ($\overline{M}_n/\overline{M}_w \simeq 2.7$). Furthermore, the molecular weight could be controlled by varying the [M]/[In] ratio (23). This suggested the presence of "living polymers." Indeed, they were demonstrated by sequential monomer addition experiments and by block copolymer synthesis. Block copolymers of (14) with acrylonitrile, styrene, and methyl acrylate were prepared with controlled block lengths.

TABLE 2.   *Summary of the Reactivity Ratios of Acrylic Monomers Which Contain Ferrocene or* η^6*-(Phenyl)tricarbonylchromium Groups.*

Organometallic Monomer)M$_1$)	M$_2$				
	Styrene	Methyl acrylate	Methyl methacrylate	Acrylo-lonitrile	Vinyl acetate
6 Solvent	EA(31)	EA(31)		EA(31)	
°C	70°	70°		70°	
r$_1$	.1	.3		.6	
r$_2$	.5	1.0		.2	
7 Solvent	Bz(32)	Bz(32)		Bz(32)	Bz(32)
°C	80°	80°		80°	80°
r$_1$	.26	.30		.34	2.0
r$_2$	1.81	.74		.74	.05
8 Solvent	Bz(34)	Bz(34)	Bz(34)		Bz(34)
°C	70°	70°	70°		70°
r$_1$	.02 (.0.3)	.14 (1x10^{-4})	.08 (.065)		1.44 (1.68)
r$_2$	2.3 (2.8)	4.46 (4.2)	2.9 (3.28)		.46 (.52)
9 Solvent	Bz(34)	Bz(34)	Bz(34)	Bz(35)	Bz(34)
°C	70°	70°	70°	80°	70°
r$_1$	.03 (4x10^{-6})	.08 (.03)	.12 (0.13)	.30	1.52 (1.36)
r$_2$	3.7 (4.1)	.82 (1.1)	3.37 (3.28)	.11	.20 (.17)

TABLE 2 (continued)

Organometallic Monomer (M_1)		M_2				
		Styrene	Methyl acrylate	Methyl methacrylate	Acrylonitrile	Vinyl acetate
14	Solvent	Bz[36]	Bz[36]			Bz[36]
	°C	70°	60°			60°
	r_1	0.41	0.76			3.4
	r_2	1.06	0.69			0.07
15	Solvent	Bz[36]		Bz[36]		Bz[36]
	°C	70°		60°		60°
	r_1	0.08		0.20		8.8
	r_2	0.58		0.65		0.06
16	Solvent	EA[37]	EA[37]			
	°C	60°	70°			
	r_1	0.10	0.56			
	r_2	0.34	0.62			
17	Solvent	EA[38]		EA[38]	EA[38]	
	°C	70°		70°	70°	
	r_1	0.04		0.90	0.07	
	r_2	1.35		1.19	0.79	

IV. ACKNOWLEDGMENT.

Support for this work was provided by the Office of Naval Research.

V. REFERENCES.

1. Pittman, Jr., C.U., Organometallic Reactions (E Becker and M. Tsutsui Eds.), Vol. 6, p. 1-62. Marcek Decker, 1977.
2. Sasaki, Y., Walker, L.L., Hurst, E. L., and Pittman, Jr., C.U., J. Polymer Sci. Chem. Ed. 11, 1213 (1973).
3. George, M.H. and Hayes, G.F., ibid. 13, 1049 (1975).
4. Pittman, Jr., C.U., Marlin, G. V., and Rounsefell, T.D., Macromolecules 6, 1 (1973).
5. Mintz, E.A., Rausch, M.D., Edwards, B.H., Sheats, J.E., Rounsefell, T.D., and Pittman, Jr., C.U., J. Organomet. Chem., 0000 (1977).
6. Pittman, Jr., C.U., Grube, P.L., Ayers, O.E., McManus, S.P., Rausch, M.D., and Moser, G.A., J. Polymer Sci. A-1, 10, 379 (1972).
7. Funita, N. and Sonogashira, K., ibid. 12, 2845 (1974).

8. Pittman, Jr., C.U. and Marlin, G.V., _ibid._ 11, 2753 (1973).
9. Pittman, Jr., C.U., Ayers, O.E., and McManus, S.P., _J._
    Macromol. Sci. Chem. A7 (8), 1563 (1973).
10. Arimoto, F.S. and Haven, Jr., A.C., _J. Am. Chem. Soc._ 77,
    6295 (1955).
11. George, M.H. and Hayes, G.F., _J. Polymer Sci. Chem. Ed._
    14, 475 (1976).
12. Aso, C., Kunitake, T., and Nakashima, T., _Macromol. Chem._
    124, 232 (1969).
13. Simionescu, C.R., et. al., _ibid._ 163, 59 (1973).
14. Rounsefell, T.D., PhD Thesis, University of Alabama (1977).
15. Pittman, Jr., C.U., et. al., _Macromolecules_ 3, 105 (1970);
    ibid. 4, 155 (1971).
16. Pittman, Jr., C.U. and Surynarayanan, B., _J. Am. Chem._
    Soc. 96, 7916 (1974).
17. Korshak, V.V., Sladkov, A.M., Luneva, L. K., and Girsho-
    vich, A.S., _Vysokomol. Soedin._ 5, 1284 (1963).
18. Ralea, R., et. al., _Rev. Roumaine Chim._ 12, 523 (1967).
19. Korshak, V.V., Dzhashi, L.V., and Sosin, S.L., _Nuova Chim._
    49 (3), 31 (1973).
20. Yurlova, G.A., Chumakov, Yu. V., Ezhova, T.M., Dzashi,
    L.V., Sosin, S.L., and Korshak, V.V., Vysokomol. Soedin.,
    Ser. A. 13 (12), 276 (1971).
21. Korshak, V.V., Dzhashi, L.V., Antipova, B.A., and Sosin,
    S.L., _Dokl. Vses. Konv. Khim Atsetilena 4th_ 3, 217 (1972).
22. Korshak, V.V., Frunze, T.M., Izynee, A.A., and Samsonova,
    V.G., _Vysokomol. Soedin. Ser. A._ 15 (3), 521 (1973).
23. Simionescu, C., Lixandru, T., Maxilu, I., and Tatrau, L.,
    Makromol Chem. 147, 69 (1971).
24. Pittman, Jr., C.U., Sasaki, Y., and Grube, P., _J. Macromol._
    Sci. Chem. A8 (5), 923 (1974).
25. Nakashima, T., Kunitake, T., and Aso, C., _Makromol. Chem._
    157, 73 (1972).
26. Pittman, Jr., C.U. and Sasaki, Y., Chem. Lett. Japan, 383
    (1975).
27. Tidwell, P.W. and Mortimer, G.A., _J. Polymer Sci._ A-1, 3,
    369 (1965).
28. Pittman, Jr., C.U. and Rounsefell, T.D., _Macromolecules_ 9,
    936 (1976) and references therein.
29. Minoura, Y. and Sakanaka, Y., _ibid._ 7, 3287 (1969); _ibid._
    4, 2757 (1966).
30. Alfrey, Jr., T., Bohrer, J.J., and Mark, H., _Copolymeriza-_
    tion, "High Polymers", Vol. 3, Interscience, New York,
    1952.
31. Pittman, Jr., C.U. and Marlin, G.V., _J. Polymer Sci. Chem._
    _Ed._ 11, 2753 (1973).
32. Pittman, Jr., C.U., Ayers, O.E., and McManus, S.P., _J._
    Macromol. Sci. Chem. A7 (8), 1563 (1973).

33. George, M.H. and Hayes, G.F., Makromol. Chem. 177, 399 (1976).
34. Lai, J.C., Rounsefell, T.D., and Pittman, Jr., C.U., Macromolecules 4, 155 (1971).
35. Ayers, O.E., McManus, S.P., and Pittman, Jr., C.U., J. Polymer Sci., Chem. Ed. 11, 1201 (1973).
36. Pittman, Jr., C.U., Voges, R.L., and Jones, W.R., Macromolecules 4, 298 (1971).
37. Pittman, Jr., C.U., Voges, R.L., and Elder, J., Macromolecules 4, 302 (1971).
38. Pittman, Jr., C.U., Ayers, O.E., and McManus, S.P., Macromolecules 7, 737 (1974).
39. Pittman, Jr., C.U. and Hirao, A., J. Polymer Sci. Chem. Ed., 0000 (1977).

PARTICIPATION OF THE FERROCENE NUCLEUS IN THE POLYMERIZATION OF VINYLFERROCENE AND ITS EFFECT ON POLYMER PROPERTIES

by
G. F. Hayes
Explosives Research and Development Establishment

and
M. H. George
Department of Chemistry
Imperial College of Science and Technology

I. ABSTRACT

It has been shown that the ferrocene nucleus participates in radical polymerizations of vinylferrocene[1]-[4] in a characteristic intramolecular electron transfer termination of the growing polymer chain radical:—

$$\sim\!\!\sim\!\!\sim\!\!CH_2-\overset{\bullet}{C}H(Fc) \longrightarrow \sim\!\!\sim\!\!\sim\!\!CH_2-\overset{-}{C}H(Fc^+)$$

where Fc denotes the ferrocenyl group $C_{10}H_9Fe$.

As a result of this reaction the polymer contains an ionically—bound high spin $(3d^5)$complex of Fe(III). This paper is concerned with a study of the effects of this paramagnetic species on the polymerization and properties of polyvinylferrocene. The complex present in the polymer appears to confer some ionic properties since the polymer precipitates from benzene during the polymerization reaction. A study of the rate of polymerization in the presence of the precipitated phase indicates that there is no autoacceleration. As the percentage of conversion increases, an increasing fraction of the polymer becomes insoluble in benzene and the majority of solvents. However, the insolubility is not due to crosslinking, as suggested previously, since the polymers are readily soluble in chloroform, and is more likely due to aggregation phenomena.

13

II.   INTRODUCTION

A recent study of the kinetics of radical polymerization of vinylferrocene has shown that the ferrocene nucleus participates in the addition reaction at the olefinic double bond. For example in benzene(1) the growing polyvinylferrocene chain radical is terminated by an electron transfer from the iron atom as depicted below:—

$$\cdots CH_2-\overset{\bullet}{CH} \quad \xrightarrow{k_t} \quad \cdots CH_2-\overset{-}{CH}$$

This reaction is monomolecular and intramolecular with respect to the chain radical. The resultant ferricinium cation produced in this termination step does not persist in the polymer but rearranges to form a high spin $(3d^5)$ ionically bound complex of $Fe(III)$ which confers paramagnetism to the polymer.(2) Because of this termination step, radical polymerization of vinylferrocene does not follow the classical kinetics of vinyl polymerizations and the rate of polymerization is first order in both monomer and initiator concentration. The rate of polymerization is given by:—

$$R_p = (2k_p k_d f/k_t) \, [M] \, [In]$$

whilst the degree of polymerization is independent of the initiator concentration.

Although the ease with which the ferrocene nucleus is oxidized often limits the applications of ferrocene compounds, this reaction can be used to advantage with vinylferrocene since this monomer will undergo radical polymerization simply by heating in halogenated solvents.(3) In these solvents, ferrocene derivatives are oxidized in a characteristic redox reaction and in the case of vinylferrocene the concomitant radicals produced will initiate radical addition at the double bond to produce low molecular weight polymers.

$$CH_2{=}CH\text{-}Fe + RX \longrightarrow CH_2{=}CH\text{-}Fe^+ X^- + R^{\bullet} \longrightarrow polymer$$

These polymers also undergo a monomolecular termination reaction since Mössbauer and ESR spectroscopy confirm the presence of an Fe III species. Similarly copolymers of vinyl-ferrocene and butadiene contain a paramagnetic species[4] whilst copolymerizations with fumaronitrile and diethyl fumarate show a first order dependence between the rate of copolymerization and the initiator concentration.[5]   These results suggest that even in copolymerizations, intramolecular electron transfer termination is the characteristic termination step. Some preliminary studies of the anionic polymerisation of vinylferrocene[6] have also suggested that electron transfer from the nucleus may occur, although in this system massive decomposition of the monomer tends to mask the polymerization reactions. Since participation of the nucleus appears general, it might be expected that the paramagnetic species produced may influence the polymerization in other ways. Such effects have been observed and this paper records the results of a study of radical polymerizations of vinyl-ferrocene to high conversions in benzene and in bulk.

III.   EXPERIMENTAL

All polymerizations were carried out using 2,2' azobisi-sobutyronitrile (AIBN) as initiator in sealed glass vessels at 60°C in the absence of oxygen as described previously.[1]-[4] Reaction times were varied from 10 to 240 hours and the polymers isolated and purified as before. In some reactions, a benzene-insoluble fraction was obtained which was separated from the benzene-soluble fraction by soxhlet extraction and then dried to constant weight.

IV.   RESULTS AND DISCUSSION

Some typical results of the polymerization reactions are given in Table 1. In general, the iron contents were lower than polyvinylferrocene isolated at low conversions[1] which might indicate that some decomposition can occur during extended reactions (>40 hours) particularly in bulk polymeri-zations. The molecular weights were low which is typical for vinylferrocene polymers because of the monomolecular termina-tion and the high transfer activity of this monomer.[1] [2] The polymers gave characteristic broad NMR spectra and ESR spectroscopy confirmed the presence of a paramagnetic species although the signals were weaker than from polymers isolated at low conversions in previous studies. As might be expected, termination of the chains by transfer reactions will become more important at higher conversions and will compete with

electron transfer termination, thus reducing the effective
concentration of the Fe(III) terminal unit.

TABLE 1

Bulk and Solution Polymerization of
Vinylferrocene at $60^{\circ}C$ in the Presence of AIBN

$\dfrac{[M]_o}{mol\ dm^{-3}}$	$\dfrac{[In]_o \times 10^2}{mol\ dm^{-3}}$	Conversion %	Iron Content %	$\overline{M}_n$ (GPC)[a]	$\overline{M}_n$ (VPO)[b]
1.0	1.0	9.9	23.4, 23.1	8,200	5,900
1.0	1.0	18.5	23.4, 23.1	8,300	
1.0	1.0	23.4	23.4, 23.6		6,400
4.0	1.5	12.5			12,600
"	2.0	20.0	23.6, 23.7		19,300
"	"	32.3 [c]		7,020	28,000
"	"	42.7 [c]		4,300	
"	"	54.9 [c]		8,100	
5.9	1.0	23.0 [c]	22.4, 22.5	1,400	11,200
"	4.0	40.8 [c]		24,000	
"	8.0	52.3 [c]			

a At ambient temperature in THF on benzene—soluble fraction
   only.
b Determined at $37^{\circ}C$ in benzene on benzene—soluble fraction
   only.
c Includes benzene—insoluble fraction.

IV A. RATE OF POLYMERIZATION

The rate of polymerization is linear to about 20% conversions in reactions in benzene with $[M]_o$ = 4.0 mol dm^{-3} as shown
in Figure 1. The rate then progressively decreased with time
with no evidence of autoacceleration even though the viscosity
of the reaction mixes increased throughout the reaction
eventually setting solid in reactions with high monomer concentrations. In normal vinyl polmerizations, the chain radicals
undergo a bimolecular termination step which is sensitive to
diffusion control and decreases rapidly in viscous solutions
causing the well known 'gell effect'. In vinylferrocene
polymerizations, the termination step is monomolecular and
should be relatively insensitive to the viscosity of the
reaction media and absence of autoacceleration is not

unexpected. Although the increased viscosity may not
influence the termination rate, it may slow down the propagation rate causing the observed decrease in the overall rate.

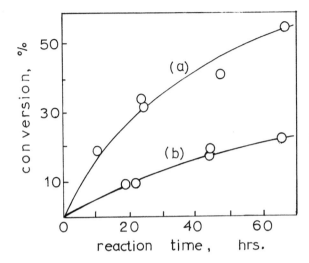

Fig 1   Percent
conversion as a
function of time
for the polymeriza-
tion of vinylferro-
cene in benzene (a)
$[M]_0$ = 4.0 mol dm^{-3}
(b) $[M]_0$ = 1.0 mol
dm^{-3}

However, since $Rp \propto [M][In]$ the decreased rate may simply
be due to a decrease in the concentration of the reactants.
It has been shown experimentally[7] that the rate and yield
can be increased by sequential or continuous addition of
initiator which supports this conclusion.

IV B. MOLECULAR WEIGHT DISTRIBUTION

As the percentage conversion increased, the molecular
weight distribution became broader and often bimodal as
reported previously.[2][7] An interesting feature in some GPC
traces of benzene–soluble fractions isolated at conversions
above 33% was the virtual absence of the usual high molecular
weight tail. A typical trace is shown in Figure 2. Since the
monomer has a high transfer activity, the polymer also
probably has a high transfer activity and therefore, at high
concentrations of polymer, i.e. high conversions, may be very
effective in limiting the molecular weight. Similar traces
have been observed from polyphenylacetylene.[8] Like polyvinylferrocene, this polymer also undergoes an intramolecular
chain termination reaction which might suggest that the main
reason for the absence of high molecular weight is the

monomolecular termination reaction. However, a concise interpretation of the GPC traces is difficult since the benzene-insoluble fraction has been ignored.

Fig 2   Gel Permeation Chromatogram of benzene-soluble polyvinylferrocene isolated at 42.7% conversion

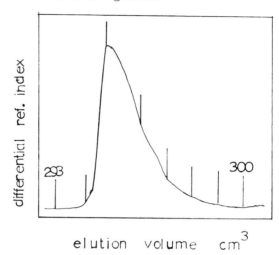

IV C. HETEROGENEOUS POLYMERIZATION AT LOW MONOMER CONCENTRATIONS

At low monomer concentrations, $[M]_0 = 1.0$ mol dm^{-3}, the clear benzene solutions became cloudy after 24 hours reaction and further reaction resulted in the formation of a yellow precipitate, the amount of which increased with time. Normally, in vinyl polymerizations in which the polymer is precipitated, a high proportion of the chain radicals are trapped and continue propagation without undergoing termination, causing a rapid and often violent increase in the rate of polymerization. This does not occur with vinylferrocene (see lower curve Figure 1) suggesting that the growing chain radicals are not precipitated from the liquid phase. Precipitation only occurred in the reactions at low monomer concentrations and is apparently caused by some polymer-solvent interaction. Mössbauer spectroscopy has indicated that the Fe(III) species present in the polymer has some ionic character which probably imparts polar properties to the polymer causing precipitation during reactions with high benzene concentrations. Since the Fe(III) species is only formed after termination, the precipitated polymer does not trap growing radicals and therefore the presence of the solid phase does not affect the rate of reaction. Surprisingly the polymers isolated from these reactions were soluble in benzene.

IV D. SOLUBILITY OF POLYMER ISOLATED AT HIGH CONVERSIONS

Polyvinylferrocene isolated at low conversions is completely soluble in benzene but an increasing fraction becomes insoluble in benzene as the percentage conversion of monomer to polymer increases. The results plotted in Figure 3 sugeest that all the polymer will be insoluble in benzene at conversions above 90%.

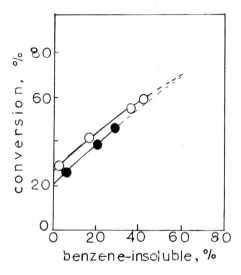

Fig 3 The effect of conversion on the amount of benzene-insoluble polymer isolated from the polymerization of vinyl-ferrocene; (O) solution polymerization, (●) bulk polymerization.

Previous workers(7) have reported that the benzene-insoluble polymer isolated from polymerizations of vinyl-ferrocene to high conversions is insoluble in all organic solvents and postulated a crosslinked structure for this insoluble fraction. Whilst vinylferrocene and its polymer readily undergo transfer reactions, these alone will not cause crosslinking. To clarify this point the solubility of the benzene-insoluble fraction was checked in a number of solvents and the results are recorded in Table 2.

TABLE 2

Solubility of Benzene–insoluble Fractions Isolated
from the Polymerization of Vinylferrocene to High Conversions

Solvent	Solubility Parameter $(cal/cm^3)^{0.5}$	Solubility
Carbon tetrachloride	8.6	insoluble
Toluene	8.9	"
THF	9.1	"
Benzene	9.2	"
Chloroform	9.3	readily soluble
1,4 Dioxan	10.0	insoluble
Carbon disulphide	10.0	slightly soluble
DMSO	12.0	insoluble
DMF	12.1	"

The benzene–insoluble fraction is indeed insoluble in
the majority of solvents. However, when placed in chloroform
the polymer dissolves immediately to give low viscosity crystal
clear solutions. Solubility cannot be readily explained on
the basis of solubility parameters; for example benzene and
chloroform have very similar values, nor does it seem to be
dependent upon solvent polarity. The intrinsic viscosities of
the benzene–soluble and benzene–insoluble fraction (determined
in fresh chloroform solutions) were 0.063 and 0.068 100 cm^3
g^{-1} respectively indicating both fractions have similar
molecular weight.

IV E. STRUCTURE OF BENZENE–INSOLUBLE FRACTION

The infrared spectra of the benzene–soluble and benzene–
insoluble fractions, examined in KBr discs, are shown in
Figure 4, and are essentially identical.

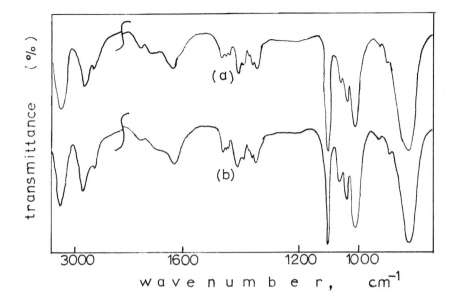

Fig 4   Infrared spectra of (a) benzene—soluble and
         (b) benzene—insoluble fractions of polyvinyl-
         ferrocene isolated at 42.7% conversion ,
         examined in KBr discs.

         Unlike the spectra of polymer isolated at low conversions,
both contain rather broad absorptions near 1630 cm^{-1} probably
due to unsaturation introduced by transfer reactions.[1][2]
Elemental analysis of the two fractions also revealed no
differences. Iron contents recorded in Table 3, in general,
were between 22% to 24%. Similarly the UV spectra of the two
fractions, determined in chloroform, were identical. The
molar absorptivities are in general agreement as shown in
Table 4. The slight discrepancy may be due to the use of a
repeat unit based on $\{CH_2-CHFc\}_n$ since other units may be
present at high conversions[2]. Care was taken to record the
spectra of fresh solutions since the polymer is oxidized on
standing in chloroform.

TABLE 3

Iron Contents of Benzene–Soluble
and Benzene–Insoluble Polyvinylferrocene

$[M]_o$ mol dm^{-3}	$[In]_o$ x 10^2 mol dm^{-3}	Iron Content %	
		Benzene–Soluble	Benzene–Insoluble
4.0	2.0	24.1, 23.2	23.1, 22.2
4.0	2.0	22.9, 23.0	22.0, 24.0
5.9	1.0	22.4, 22.5	23.0, 22.4
5.9	4.0	23.6, 23.1	24.2, 23.7

TABLE 4

Electronic Spectra of Benzene–Soluble and Benzene–Insoluble
Fractions Isolated from Polymerization of Vinylferrocene to
High Conversions (>30%)

Fraction	λ max nm	ε dm^3 mol^{-1} cm^{-1}	Solvent
Benzene–soluble	330 446	71 95	Chloroform
Benzene–insoluble	330 448	61 96	"

These results suggest that the benzene–soluble and benzene–
insoluble fractions isolated from polymerization of vinylferrocene to high conversion have similar structure. It would
appear that insolubility is caused by aggregation phenomena.

IV F. AGGREGATION OF POLYVINYLFERROCENE IN SOLUTION

A proportion of the polymer molecules obtained by free radical polymerization of vinylferrocene in benzene will always contain an Fe(III) ionic-type complex due to the 'intralectran' termination.[1][2] It is feasible, therefore, to envisage some mutual interaction between adjacent chains. The indications are that these interactions are quite specific since chloroform is, apparently, the only solvent which breaks them down and solubilizes the polymer. The formation of benzene-insoluble polymer at high conversions seems to only occur in benzene solutions, since radical polymerization in chloroform to high conversions ($\approx$80%) does not produce benzene-insoluble polymer.[3] Further support for polymer-polymer aggregation in benzene is given by the low value of α in the Mark-Houwinck equation, ($\alpha = 0.47$), [7] since values below 0.5 are characteristic of systems where aggregation is important; for example $\alpha = 0.2$ for polyvinylchloride in dioxan.[9] If association phenomena is present, one might expect some effects on solution viscosities. Further studies have in fact shown that the viscosity of solutions of polyvinylferrocene in mixed benzene/chloroform solvents increase on dilution. (G.F. Hayes and M.H. George, Note in preparation.)

V. CONCLUSIONS

The results of the polymerization of vinylferrocene to high conversions are consistent with previous kinetic studies. (1,2) Because the termination reaction is monomolecular with respect to the chain radical, it is not diffusion controlled and therefore there is no autoacceleration of the rate of polymerization in viscous solutions. Similarly, the rate of polymerization is not increased in the presence of precipitated polymer as usually occurs in vinyl polymerizations since the polymer is only precipitated after termination and therefore does not trap growing chain radicals. As the percentage of conversion increases an increasing amount of the polymer becomes insoluble in benzene (and most other solvents) but this fraction is not cross-linked since it is readily soluble in chloroform and has the same structure as the benzene-soluble fraction. The insolubility appears to be due to aggregation phenomena, probably caused by the presence of the ionically-bound Fe(III) complex in polyvinylferrocene.

VI. REFERENCES

1   George, M.H. and Hayes, G.E, J. Polym. Sci, Polym. Chem. Ed, <u>13</u> 1049 (1975)

2   George, M.H. and Hayes, G.E, <u>ibid</u> <u>14</u> 475 (1976)

3   George, M.H. and Hayes, G.E, Polymer, <u>15</u> 397 (1974)

4   George, M.H. and Hayes, G.E, Die Makromol Chemie, <u>177</u> 399 (1976)

5   Tada, K, Higuchi, H, Yoshimura, M, Mikawa, H., and Shirota Y, J. Polym. Sci, Polym. Chem. Ed, <u>13</u> 1737 (1975)

6   Hayes, G.F. and Young, R.N, Polymer (in press)

7   Sasaki, Y, Walker, L.L, Hurst, E.L. and Pittman, Jr. C.U, J. Polym. Sci, Polym. Chem. Ed, <u>11</u> 1213 (1973)

8   Biyani, B, Campagne, A.J, Daruwaller, D, Scrivastava C.M, and Ehrlich P, J. Macromol. Sci—Chem A9 (3) 327 (1975)

9   Sakurada, L, Kagaku, <u>27</u> 49 (1972)

SYNTHESIS AND POLYMERIZATION OF SOME FERROCENE DERIVATIVES

V.V.Korshak, S.L. Sosin
Institute of Organo-Element Cŏmpounds,
Academy of Sciences of the USSR
Moscow,USSR

INTRODUCTION

In our previous investigations on the synthesis of the ferrocene-containing polymers, $\propto$ -oxyderivatives and some vinyl derivatives of ferrocene were used as starting reagents: 1,1'-di($\propto$-oxy)ethyl ferrocene (DOEF), 1,1'-di($\propto$-oxy)isopropylferrocene (DOPF), 1,1'-diisopropenylferrocene (DIPF), acryloylferrocene (AF), etc. (1-3).

In the past 5-7 years attention of the scientists working in the field, was attracted by the polymerization of the unsaturated derivatives of ferrocene, since by using these monomers it is possible to obtain polymers having high molecular weights; moreover, application of such monomers leads to the new possibilities for the incorporation of the ferrocene units into the polymer backbones, and, as a result, to the preparation of new ion-exchangers, to the improvement of the thermal and radiation stability (e.g., UV-radiation), to new routes for polymer modification etc.(4).

As starting reagents for the preparation of ferrocene-containing polymers we have choosen mono- and divinyl derivatives of ferrocene, ethyl substituted vinyl- (VF) and divinylferrocene (DVF), ferrocenyl acetylene (FA) and 2-ferrocenyl butadiene (FB). Some of these monomers were prepared in accordance with the known (unmodified or slightly modified procedures); on the other hand for the preparation of some monomers significant modifications of the known procedures were needed.

In the course of the present investigation the optimum conditions for polymerization reactions were studies. In some cases possible mechanisms were investigated and kinetic constants were determined. In addition, we have investigated the structure and properties of the ferrocene-containing polymers obtained.

RESULTS AND DISCUSSION

We have found that polymerization of DVF proceeds especially easily by cationic and radical

mechanisms. Polymerization by cationic mechanism
has led to the formation of polymers having molecular weight up to 35000 (Table 1) (5).

TABLE 1

Polymerization of 1,1'-Divinylferrocene in Benzene Solution

Catalyst	Quantity of cata-lyst weight,%	Solvent	T,$^{\circ}$C	Time of reac-tion, hours	Yield of polymer, %		
					General	solub-le	in-so-lub-le
Without catalyst	–	Benzene	80	4	0,36	–	–
Azobuty-ronitrile	1	–"–	–"–	–"–	98	8	90
$BF_3(OEt_2)$	2	–"–	–"–	1	85	45 a)	40
–"–"–"–	1	–"–	20	18	75	15	60
BuLi	1,5	Diethyl ether	0	1	13	1,7	11,3
$Al(iso-C_4H_9)_3 \cdot TiCl_4$ Al/Ti=1:1	5	Benzene	20	5	26,3	3,3	23
$Al(iso-C_4H_9)_3 \cdot TiCl_4$ Al/Ti=3:1	5	–"–	20	4	45,7	10	35,7

a) Molecular weight 35.000 (ebullioscopy in benzene).
   m.p.360-380°C (thermomechanical data).

One of the pecularities of the DVF polymerization
is the formation of the cyclolinear structures resulting from
the formation of α-ferrocenylcarbonium ions due to the intramolecular electrophilic attack of cation reactive centre on vinyl
groups.

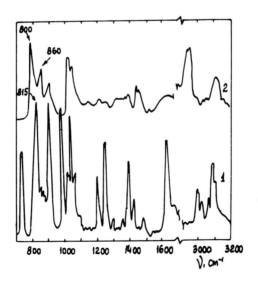

This structure was confirmed by the IR-data /splitting of the deformational vibration absorption maxima of C–H bond (815 cm^{-1}) into two maxima (800 and 860 cm^{-1})/ (6) (Fig.1).

Fig. 1.   Infrared spectra of DVF (1) and poly-DVF (2).

These results were confirmed by two Japanese scientists
(7). The thermal stability of the poly-DVF is higher than that
of the known ferrocene-containing polymers, but doesn't exceed
380-400°C in an inert atmosphere (Fig. 2).

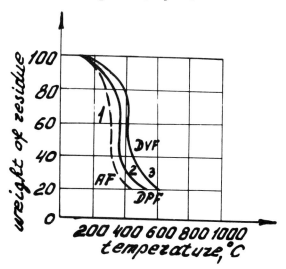

**Fig.2. Thermal stability of polymers from:
1) AF, 2) DIPF, 3) DVF**

In accordance with "polymer different-unit"
conception (8), developed by Prof. V.V. Korshak, an
introduction of methylene and ester groups (via
the polyrecombination method) (15-16) into the macromolecules of the ferrocene-containing polymers
between ferrocene nucle i (17) leads to the decrease of their thermal stability from 400 to 300°C
comparable to the thermal stability of "pure" ferrocene and the low molecular weight polyferroceny-
lene obtained via mercury derivatives of ferrocene. Probably the same defect (uncyclized) units
are present in poly-DVF.

Other polymers obtained by us previously, e.g.
polymers containing epoxide derivatives,
possessed even lower thermal stability - at about
300°C. As a result, the possibilities of application of ferrocene-containing polymers as thermally stable materials are doubtful as it was mentioned in Larkovskiy's review (18).

Kinetics of VF and DVF homopolymerization by
radical mechanism was investigated using dilato-

metric and gravimetric methods.   In the course of
the investigation rubber-like copolymers of VF and
isoprene were obtained (9); Rubber-like copolymers
containing up to 23% of Fe were obtained.

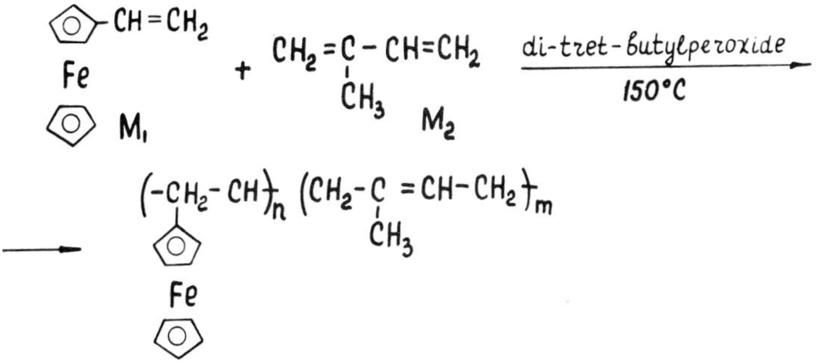

$$
\begin{array}{c}
\underset{\substack{\text{Fe}\\\text{M}_1}}{\overset{\text{CH}=\text{CH}_2}{\bigcirc}}
+ \;
\underset{\text{M}_2}{\overset{\displaystyle CH_2=\underset{\underset{CH_3}{|}}{C}-CH=CH_2}{}}
\quad \xrightarrow{\substack{\text{di-tret-butylperoxide}\\150°C}}
\end{array}
$$

$$
\longrightarrow \quad
\left(-CH_2-\underset{\bigcirc}{CH}\right)_{\!n}\left(CH_2-\underset{\underset{CH_3}{|}}{C}=CH-CH_2\right)_{\!m}
$$

$$
\begin{array}{c}
\text{Fe}\\
\bigcirc
\end{array}
$$

$$
z_1=0.54 \qquad z_2=1.56 \qquad Q_1=1.288 \qquad Q_2=3.33 \qquad e_1=-0.805
$$

$$
e_2 = -1.22
$$

The rate constants of the copolymerization of VF
with isoprene (9) were determined by the Mayo-Le-
wis method.
At the next stage of the investigation the
polymerixation of FA was studied.   Treatment of the
monomer with di-tert. butyl peroxide led (via the
stage of dehydrodimer) to the formation of ladder
polymer containing a system of conjugated bonds
(10).
Another way for the preparation of this polymer (using metallic sodium, via sodium ferrocenyl-
acetylenide) also has led to the formation of the
same polymer having intensive ESR signal.

$$Fc-C\equiv CH \xrightarrow{Na} Fc-C\equiv C^{\ominus}Na^{\oplus} + HC\equiv \underset{Fc}{C} \xrightarrow{-NaH}$$

$$\longrightarrow \underset{Fc}{C}\equiv C - \underset{Fc}{C}\equiv C \longrightarrow$$

$$Fc = Fe \underset{\bigcirc}{\overset{\bigcirc}{}}$$

FA polymerization was carried out by other investigators using other initiators, e.g. using benzoyl peroxide (19). An intermediate product formed during the FA synthesis - - chloroformylvinylferrocene - readily undergoes polymerization, with the formation of highly conjugated polymer possessing semiconducting properties (20):

Investigation of the copolymerization of the FA with other monomers (e.g., with chloroprene) led

to an explosive reaction. Investigation of the influence of
the degree of unsaturation of the substituents on the polymerization procedure has shown that ferrocene itself inhibits the simultaneous polymerization of
chloroprene up to high temperature 100°C) (11).

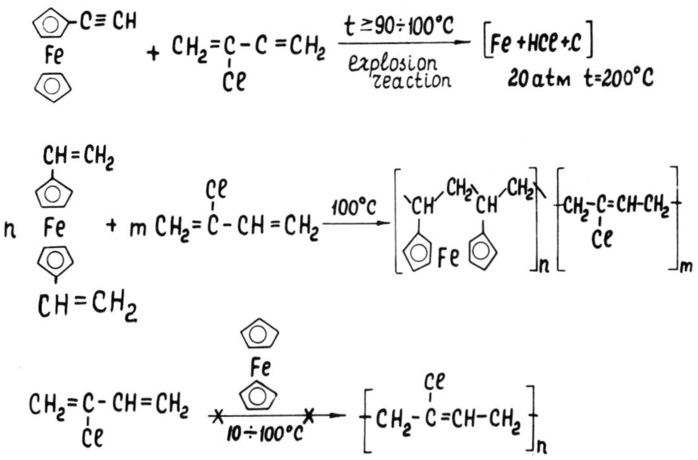

$$\underset{Fe}{\text{(ferrocenyl)}}-C\equiv CH \; + \; CH_2=C-C=CH_2 \; \underset{\substack{\text{explosion} \\ \text{reaction}}}{\xrightarrow{t\geq90\div100°C}} \; \left[Fe+HCl+C\right] \atop 20\,atm \; t=200°C$$
$$\overset{|}{Cl}$$

$$n \; \underset{CH=CH_2}{\overset{CH=CH_2}{\underset{Fe}{\bigcirc}}} \; + \; m \; CH_2=\overset{Cl}{\underset{|}{C}}-CH=CH_2 \; \xrightarrow{100°C} \; \left[\overset{CH_2}{\overset{\diagup}{CH}}\overset{CH_2}{\overset{\diagdown}{CH}} \atop \underset{Fe}{\bigcirc} \bigcirc \right]_n \left[CH_2-\overset{Cl}{\underset{|}{C}}=CH-CH_2\right]_m$$

$$CH_2=\overset{Cl}{\underset{|}{C}}-CH=CH_2 \; \underset{10\div100°C}{\overset{\underset{Fe}{\bigcirc}\bigcirc}{\xrightarrow{\quad x \quad}}} \; \left[CH_2-\overset{Cl}{\underset{|}{C}}=CH-CH_2\right]_n$$

We further investigated the synthesis and polymerization of the VF and DVF ethyl derivatives
prepared by the following scheme (12) (*Table 2*):

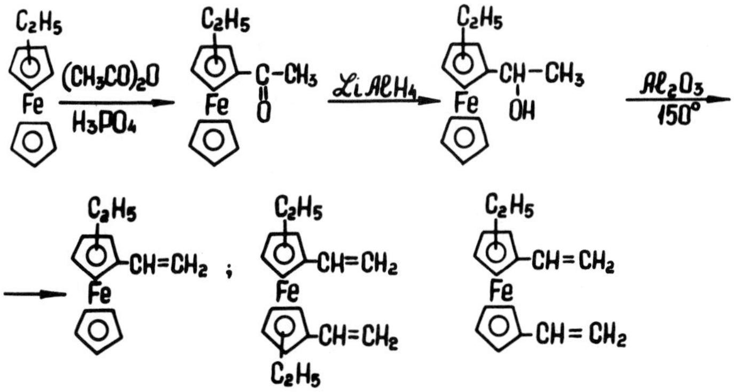

$$\underset{Fe}{\overset{C_2H_5}{\bigcirc}} \xrightarrow[H_3PO_4]{(CH_3CO)_2O} \underset{Fe}{\overset{C_2H_5}{\bigcirc}}-\overset{O}{\underset{}{C}}-CH_3 \xrightarrow{LiAlH_4} \underset{Fe}{\overset{C_2H_5}{\bigcirc}}-\overset{}{\underset{OH}{CH}}-CH_3 \xrightarrow[150°]{Al_2O_3}$$

$$\xrightarrow{\quad} \underset{Fe}{\overset{C_2H_5}{\bigcirc}}-CH=CH_2 \; ; \quad \underset{Fe}{\overset{C_2H_5}{\bigcirc}}-CH=CH_2 \quad \underset{Fe}{\overset{C_2H_5}{\bigcirc}}-CH=CH_2$$
$$\underset{C_2H_5}{\bigcirc}-CH=CH_2 \quad \bigcirc-CH=CH_2$$

I EVF   II DEDVF   III δDVF

TABLE 2

Data of Elementary Analysis and B.p. for Monomers

Monomer	B.p.,(^{o}C)	Data of elementary analysis					
		Found,%			Calculated,%		
		C	H	Fe	C	H	Fe
EVF $C_{14}H_{16}Fe$	76–81/7·10^{-4} mm of Hg	68,95	6,60	23,26	69,98	6,71	23,31
$C_{18}^{DEDVF}H_{22}Fe$	85–110/5·10^{-4} mm of Hg	73,31 73,18	7,60 7,59	18,82 18,93	73,48	7,54	18,98
EDVF $C_{16}H_{18}Fe$	–	71,81 71,84	6,83 6,86	20,80 20,75	72,20	6,82	20,98

In fact there was obtained not pure homo-
annular EVF, but a mixture of homo- and hetero-
annual isomers. Results of these derivatives' polymerization were compared with the results of VF
and DVF polymerizations under the same conditions
studied by us previously. It was found that the
presence of an ethyl group in the cyclopentadienyl ring
of VF and DVF decreases the ability for polymerization of these compounds by any of the mechanisms.

The most active among these monomers seems to
be the EDVF; cationic polymerization of this monomer leads to the formation of soluble polymer with
higher yield (72%) but lower molecular weight
(8400) of the polymer in comparison with the polymerization of DVF (45%, 35000) as seen from Table 3.
During the polymerization of other ethyl derivatives the formation of dimers (cyclodimers) (by cationic mechanism) seems to be the compete reaction.

In the synthesis and polymerization of FB (13), we
worked out preparative method for the synthesis of FB-
different from the published procedure (21). The de-
hydratation of carbinol was carried out by a saturated
solution of Na.HSO_3, instead of methylchlorformate.
Yield of the product reached 90%.

The investigation of this monomer polymerization was carried out first. During the polymerization by cationic mechanism competing reactions –
cyclodimerization and the formation of ladder poly-

TABLE 3

Polymerization Results of Monomers

Monomer	Radical polymerization				Cationic polymerization				Anionic Polymerization			
	Yield,%	Mol. weight	Reaction conditions Solvent	Time, h	Yield, %	Mol. weight	Reaction conditions Solvent	Time h	Yield,%	Mol. weight	Reaction conditions Solvent	Time, h
VF	68	17200	–	13	56	795	benzene	9	– 1)	–	–	–
EVF	56	6770	–	13	60	480	benzene	9	– 1)	–	–	–
DVF	98	–	benzene	4	45	35.000	–	–	13	–	–	–
DEDVF	45	1900	–	3	95	640	benzene	1	33,3	1080	benzene	3
EDVF	31,8	5200	benzene	10	72,7	8.100	benzene	1	– 1)	–		–

1) Unreacted initial monomer; Ability to polymerization
EDVF DEDVF EVF.

mers (more truly oligomers of molecular weight up
to 2000)-take place (13) *(Table 4):*
   Cationic Polymerization of 2-Ferrocenylbutadiene:

$$CH_2=CH-C=CH_2 \quad \xrightarrow{H^{\oplus}} \quad CH_3-C^{\oplus} \quad \xrightarrow{\text{I}}$$

$$\text{(with } Fc \text{ substituent, } CH=CH_2 \text{ group)}$$

$$\longrightarrow CH_3-C-CH_2-C^{\oplus} \longrightarrow$$

$$Fc = \quad \underset{\text{Fe}}{\overset{\bigcirc}{\bigcirc}}$$

Cyclodimerization:

$$CH_2=CH-\underset{\overset{|}{Fe}}{\overset{CH_3}{C}}-CH_2-\underset{\oplus}{C}-CH=CH_2 \longrightarrow$$

$$CH_2=CH-\underset{\overset{|}{Fe}}{\overset{CH_3}{C}}-CH_2$$
$$\qquad\qquad\qquad C-CH=CH_2$$

TABLE 4

Polymerization of 2-Ferrocenylbutadiene

VB gr	Polimerization conditions			Yield,%		Notes
	Catalyst	T,°C	Time, hours	Oligo-mer	Dimer	
9,0	Azobutiro-nitryle	80	20	–	80	Blok polymeri-zation
9,6	BF$_3$·OEt$_2$	20	2	trace	72	
8,5	BF$_3$·OEt$_2$	60	3	43	34	Mol.weight of oligomer 2000
9,5	TiCl$_4$-Al(iso-Bu)$_3$	20	3	trace	93	
10,0	TiCl$_4$-Al(iso-Bu)$_3$	60	3	49	50	Mol.weight of oligomer 1400
10,0	Na-naphtha-lene	20	3	–	90	Polymerization in TgF solution
7,0	BF$_3$·OEt$_2$	60	6	–	67	Polymerization of dimer. Unreacted mono-mer.

And synthesis and polymerization of ferrocene di-epoxide derivatives.

The polymerizability of the ferrocene diepoxi-de derivatives hasn't been studies before. The synthesis of the ferrocenedicarboxylic acid digly-cidyl ester DEFK was carried out by the interacti-on of ferrocenedicarboxylic acid-1,1'-dichloranhyd-ride and glycidol; yield of product was 91% (14). DGF was prepared in accordance with the procedure described by Japenese authors (22):

$$CH_2\text{-}CH\text{-}CH_2O\text{-}\overset{O}{\underset{}{C}}\text{-}Fe\text{-}\overset{O}{\underset{}{C}}\text{-}OCH_2\text{-}CH\text{-}CH_2 \qquad DEFK \quad (14)$$

$$CH_2\text{-}CH\text{-}CH_2\text{-}Fe\text{-}CH_2\text{-}CH\text{-}CH_2 \qquad DGF \quad (22)$$

With DEFK and DGF some oligomeric products were obtained with phthalic anhydride and
carborane derivatives and different bifunctional compounds are given in Table 5.

TABLE 5

Interaction of DGF and DEFK with Different
monomers [c]

Di epoxy diriv.	Second component	Reaction conditions		Soluble product	
		Time, h	T, °C	Yield, %	Mol.weight (ebullio-scopy)
	Phthalic anhydride	1,0	50	30[a]	650 (cryoscopy in benzene)
DVF	o-Dimethylolcarbo-rane	6,0	100	96,5	540
	m-Carboranecarboxy-lic acid	0,3	20	99	1250
DEFK	n-Carboranecarboxy-lic acid	0,3	20	99	720
	Phthalic anhydride	8,0	80	29	2400
	o-Dimethylol car-borane	2,0	130	91	1600
	-"-"-	4,0	80	80	-
	Phthalic acid	1,0 2,0	50 100	37 47,5	500
	Dian	6,0	100	100 [b]	-
	-	6,0 4,0	100 80	38 [b]	-

a) Precipitated from benzene into hexane
b) Copolymer becomes insoluble on heating
c) The structure of oligomers is confirmed by IR-spectra and
   elementary analysis data.

The generalities discovered during the investigation polymerization of vinyl- and divinyl derivatives
are attributed to: 1) tendency for the formation of stable
ferrocenylcarbonium ions and, as a result, 2) preferable
polymerization by cationic mechanism, 3) easy formation of
cyclolinear and cyclochain ladder structures, the competing
reaction being the formation of cyclodimers.

ACKNOWLEDGMENTS
     Our coworkers Antipova B.A., Jashi L.V.,
Alekseeva, V.P., Ezhova, T.M., Litvinova, M.D., Afo-
nina, R.I. took part in these investigations.

REFERENCES
(1) Sosin S.L., Korshak V.V., Frunze T.M., Tverdo-
     hlebova I.I., Dokladi Akad.Nauk SSSR, v.175,
     N 5,1976 (1967).
(2) Sosin S.L., Korshak V.V., Frunze T.M., Dokladi
     Akad.Nauk SSSR, v.179,N 5,1124 (1968).
(3) Sosin S.L., Galkina M.D., Korshak V.V., Vysoko-
     mol.Soedin., XVB,N 5,373 (1973).
(4) Pittman C.U., Voges C.U., Jones W.R., Macromo-
     lecules, 4,291 (1971)
(5) Sosin S.L., Jashi L.V., Antipova B.A., Korshak
     V.V., Vysokomol.Soedin., v.XII, NO 9, 699 (1970).
(6) Sosin S.L., Jashi L.V., Antipova B.A., Korshak
     V.V., Vysokomol.Soedin., v.XVIB, NO 5, 347 (1974)
(7) Kunitake T., Nakashima T., Aso C., Makromolek.
     Chemie, 146,79 (1971).
(8) Korshak V.V., "Different unit in polymers",
     Science, Moscow,1977.
(9) Jashi L.V., Sosin S.L., Antipova B.A., Korshak
     V.V., Vysokomol. Soedin., XBIB, NO. 6, 427 (1974).
(10) Kirshak V.V., Jashi L.V., Sosin S.L., Inter-
     national Symposium of macromol.chemistry in
     Hèlsinki, v.2,829 (1972).
(11) Korshak V.V., Sosin S.L., Jashi L.V., Antipo-
     va B.A., Vysokomol.Soedin., XIV, NO # (1972).
(12) Sosin S.L., Antipova B.A., Ezhova T.M., Jashi
     L.V., Korshak V.V., Vysokomol.Soedin., v.XVIIIA
     NO 1, 34 (1976).
(13) Korshak V.V., Sosin S.L., Alekseeva V.P., Afo-
     nina R.I., Vysokomol.Soedin., XVIIB, NO 10, 779
     (1975).
(14) Sosin S.L., Alekseeva V.P., Litvinova M.D.,
     Korshak V.V., Zhigach A.F., Vysokomol.Soedin.
     v.XVIIB,N) 9, 703 (1976).
(15) Korshak V.V., Sosin S.L., Alekseeva V.P., J.
     Polymer Sci., v. 52,213 (1961).
(16) Korshak V.V., Sosin S.L., Policombinazione
     reazione, di. "Enciclopedia internazionale di
     chimica",v.VII.1033.
(17) Rosenberg H., Neuse E.W., J.Organomet.Chem.,
     6,NO. 1, 76 (1966).
(18) Larkovsky X.D., Vysokomol.Soedin., v.XVA,NO 2,
     314 (1973).

(19) Simonesku Cr., Lixandru T., Dimitresku S.,
     39 Congress international de chimie indust-
     riale Bucharest, p.190 (1970).
(20) Jurlova G.A., Chumakov J.V., Ezhova T.M.,
     Jashi L.V., Sosin S.L., Korshak V.V., Vyso-
     komolek. Soedin., v.XIIIA, NO 12, 2761 (1971).
(21) Navak R.W., Stevens T.S., Howard M., J. Or-
     ganic Chem., 36,1699 (1971).
(22) Watanabe H., Motoyama J., Hàta K., Bull.Chem.
     Soc.Japan, 39,784 (1966).

REAL AND PROPOSED ORGANOMETALLIC π-MONOMERS

D.W. Slocum
Southern Illinois University
M.D. Rausch
University of Massachusetts
A. Siegel
Indiana State University

The classes of organometallic coumpounds which can be used in polymerization reactions are limited to those monomers which:

1. Contain the organometallic moiety only as a pendant group, i.e., the organometallic complex is not part of the polymer skeleton.

2. Are bonded as an integral part of the polymer skeleton. The type of polymers found in this second category must either a) be formed from a polyfunctional organometallic compound or b) become a polyderivatized organometallic during the course of polymerization.

Mono-substituted organometallic π-monomers typified by vinyl complexes provide examples of classification 1 while di- and poly-substituted organometallic π-monomers typified by divinyl complexes provide examples of classification 2. These possibilities are summarized in Figure 1.

Pendant side-chain and both types of skeleton organometallic polymers are now known. The majority of the organometallic monomers utilized to this point have either been unsubstituted or monosubstituted derivatives and of these only the ferrocene system has been studied extensively. Certain disubstituted compounds have also been examined.

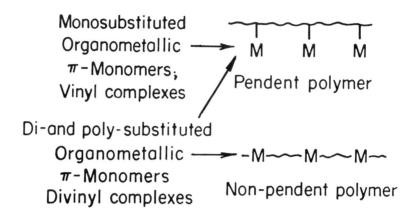

Fig. 1. Classification of polymers which can be obtained from vinyl or divinyl π-monomers.

Recent advances in synthetic organometallic chemistry have afforded new routes to suitably derivatized compounds which are available for study for the first time as well as improved methods for the synthesis of known mono- and poly-substituted complexes. A variety of unsaturated derivatives of several transition metal complexes are now represented. Unsaturated derivatives of ferrocene which have been used as monomers are illustrated in Figure 2. These include vinyl-ferrocene (1) and isopropenylferrocene (2), acryloylferro-cene (3) and trans cinnamoylferrocene (4), 1-ferrocenyl-1, 3-butadiene (4,5), 1,1'-divinylferrocene (6) and 1,1'-diiso-propenylferrocene (7).

Acylate π-monomers are summarized in Figure 3. These include ferrocenylmethyl acrylate (8), ferrocenylmethyl methacrylate (9), ferrocenylethyl acrylate (10,11) and ferrocenylethyl methacrylate (10,11). Chromium monomers are represented by η^6-(benzyl acrylate)- (12), η^6-(2-phenyl-ethyl acrylate)- (12b,13) and η^6-(2-phenethyl methacrylate) chromium tricarbonyl (14). Lastly, η^4-(2,4-hexadiene-1-yl acrylate)iron tricarbonyl (15) has been utilized.

Other vinyl organometallic π-monomers are depicted in Figure 4. Styrenechromium tricarbonyl (16), vinylcyclo-pentadienylmanganese tricarbonyl (17) and η^4-(1,3,5-hexa-trienyl)iron tricarbonyl (18) are well known. Of more recent interest is the series of platinum and palladium deriva-tives of styrene prepared by Fujita (18).

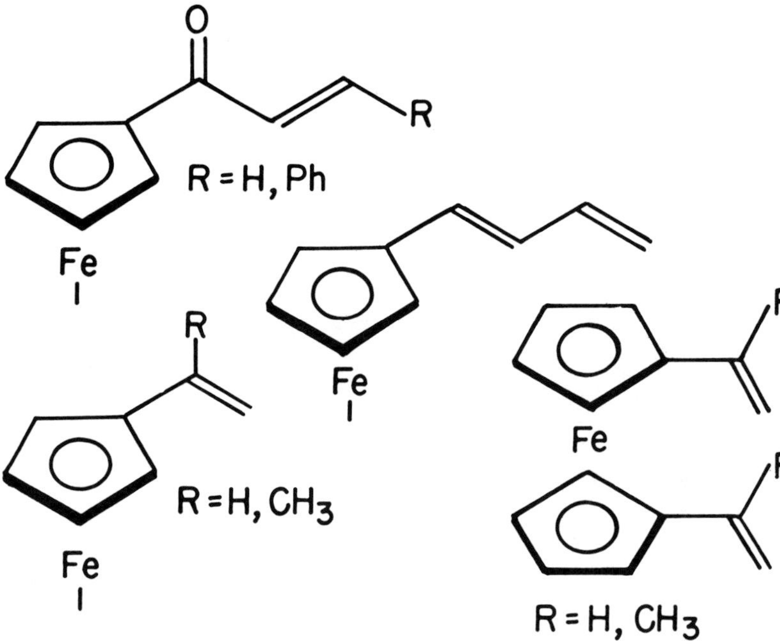

Fig. 2. *Known unsaturated ferrocene-containing monomers.*

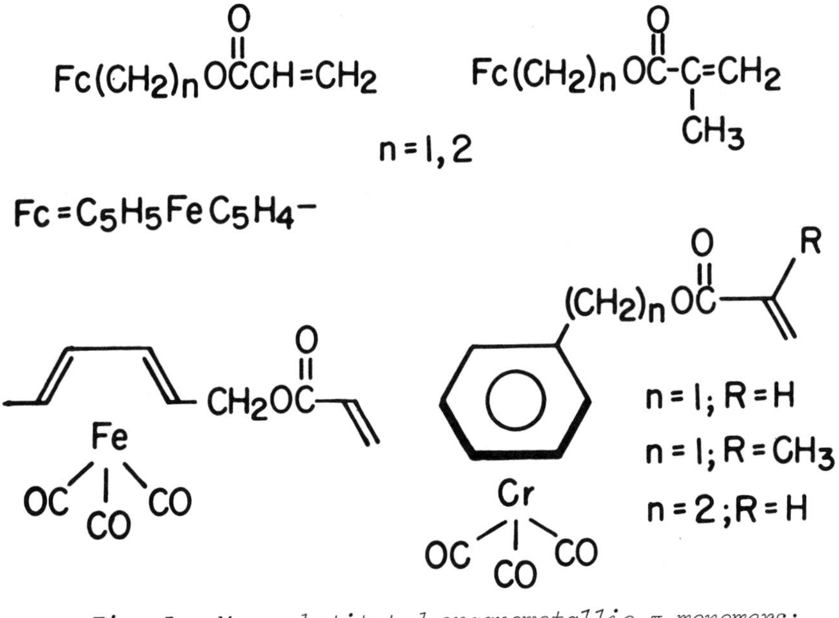

Fig. 3. *Monosubstituted organometallic π-monomers; acrylates.*

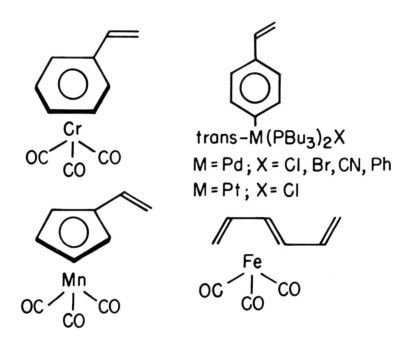

Fig. 4. Other vinyl organometallic π-monomers.

The first vinylmetallocene to be reported was vinylferrocene prepared in low yield by the distillation pyrolysis
of 1-ferrocenylethanol (la). Pyrolysis of the corresponding
acetate also produced vinylferrocene in moderate yields (lb).
Dehydrohalogenation of various haloferrocenes has since
been described and extended to the synthesis of substituted
vinylmetallocenes (lc,20). These transformations are summarized in equation 1.

$$\underset{X = -OH, -O_2CCH_3\ -Cl}{\overset{\overset{\text{X}}{|}}{FcCHCH_3}} \xrightarrow{\ -HX\ } \underset{\text{(low to moderate yields)}}{FcCH=CH_2} \tag{1}$$

Two of the most reliable and convenient syntheses of
vinylferrocene and related vinylmetallocenes developed in
these laboratories involve (a) modification (6) of the
pyrolysis procedure of Arimoto and Haven (la) and (b) application of the Wittig reaction (21).

Pyrolysis is carried out in a vacuum sublimer containing an intimate mixture (ca 3:1) of activated alumina and the appropriate carbinol by partial immersion in an oil bath at 135-180° (10-35 mm). Under these conditions, the vinylmetallocene sublimes readily onto the coldfinger to the exclusion of unpyrolyzed unsublimed carbinol. The overall yield of vinylmetallocene is about 50% based on ferrocene as the starting material. These overall transformations are illustrated in equation 2 for the preparation of vinylferrocene. Substituted vinylferrocenes may also be prepared by this procedure as illustrated in equation 3 and listed in Table 1.

$$FcH \xrightarrow[BF_3]{Ac_2O} FcCOCH_3 \xrightarrow[MeOH/H_2O]{NaBH_4} Fc\underset{OH}{C}HCH_3$$

$$\xrightarrow[\substack{alumina \\ 140-180°, 8h \\ 10-30 \ mm}]{vacuum \ sublimer} FcCH=CH_2 \qquad 60-90\%$$

(2)

$$Fc\underset{R}{\overset{OH}{C}}-CH_2R' \xrightarrow[\substack{alumina, \ 8h \\ 140-180° \\ 12-25mm}]{Vacuum \ sublimer} FcCR=CHR'$$

(3)

TABLE 1

Monosubstituted Vinylferrocenes (FcCR=CHR')
Prepared via Pyrolysis Procedure

R	R'	m.p. (°C)	% Yield
H	Ph	120-21	81
H	$2-C_4H_4S$	90-91	75
CH_3	H	67-68	82
Ph	H	29	81
H	CH_3	24-25	63

Table 2 summarizes a variety of different dehydration experiments carried out with 1-ferrocenylethanol. The conversions are high and the procedure is limited only by the size of the sublimer. The excellent yields of vinylferrocene produced under the variety of specified conditions illustrates the reliability of the method.

TABLE 2

Various Experimental Conditions for the
Dehydration of $C_5H_5FeC_5H_4CHOHCH_3$[a]

T(°C)	p(mm)	t(hr)	% Yield
147	24	5	85
135	10	7	86
160	13	7	86
153	35	8	81
153	20	7	86
158	11	8	89

a. Runs made on 10-20g of alcohol.

This procedure is also adaptable to the preparation of other vinylmetallocenes. Vinylruthenocene and vinylosmocene preparations are illustrated in equation 4 and the preparation of 1,1' - divinylferrocene (6) and 1,1' - divinylruthenocene is illustrated in equation 5. Both the versatility and utility of the procedure are indicated by these syntheses.

The pyrolysis procedure is also applicable to those vinylmetallocenes which do not sublime under these conditions. In such cases the olefin may be extracted into a suitable solvent and purified by chromatography on silica gel or alumina.

Vinylferrocene, 3-ferrocenyl-1-propene, β-phenylvinylferrocene and vinylruthenocene appear to be reasonably stable in the crystalline state. All other vinylmetallocenes slowly darken over a period of time upon exposure to air. Organic solutions of vinylmetallocenes slowly darken over a period of time.

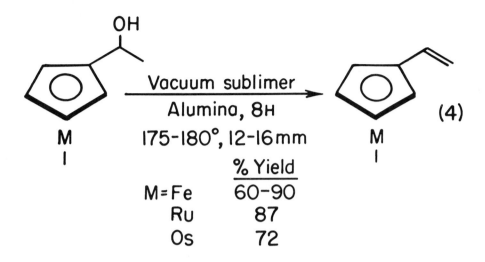

$$M=Fe \quad 60-90$$

$$Ru \quad 87$$

$$Os \quad 72$$

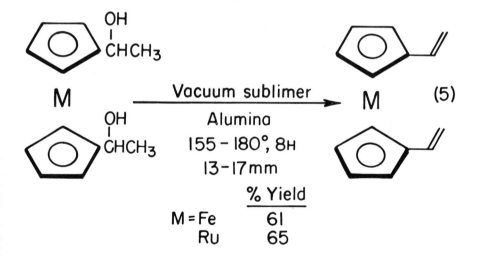

$$M=Fe \quad 61$$

$$Ru \quad 65$$

A second effective preparation of vinylmetallocenes involves application of the Wittig reaction (equation 6).

$$FcCOR_1 + Ph_3\overset{+}{P}-\overset{-}{C}R_2R_3 \xrightarrow[\substack{\text{refluxing} \\ \text{ether}}]{\text{r.t. or}} FcCR_1=CR_2R_3 \quad (6)$$

R_1, R_2, R_3 various combinations H, CH_3, Ph

In this procedure a slight excess of phosphonium salt is mixed with n-butyllithium for about 15 minutes whereupon formation of the phosphorane is complete. An appropriate organometallic carbonyl moiety is introduced as a solid or in ether solution and the mixture is refluxed for one hour. The vinylmetallocene is readily purified by column chromatography on alumina or silica gel. Table 3 lists the vinylmetallocenes produced by this method.

TABLE 3

Preparation of $FcCR_1=CR_2R_3$ <u>via</u> the Wittig Procedure

R_1	R_2	R_3	m.p. ($^\circ$c)	% Yield
H	H	H	52-53	87
H	H	Ph	100-21	75
H	H	CH_3	oil	75
H	Ph	Ph	75-76	50
Ph	H	H	27	64
CH_3	H	H	77-78	83
n-Bu	Ph	H	56-57	75

Preparation of a series of 2-substituted vinylferrocenes was the subject of a paper from the Neckers Laboratories several years ago (22). In this procedure, the directed metalation reaction (23), specifically the 2-lithiation of ferrocene, was utilized. Subsequent elimination of the elements of $HN(CH_3)_3^+$ from the quaternary ammonium salt afforded various 2-substituted vinylmetallocenes. These transformations are summarized in equation 7. Attempts to produce 1,2-divinylferrocene utilizing this procedure failed. Equation 8 illustrates a straightforward way in which this technique could be used to prepare a series of 2-substituted ferrocenylmethyl acrylates.

The first ethynylmetallocene to be described in the literature was ethynylferrocene prepared in 25% yield by bromination of vinylferrocene followed by dehydrobromination (1c). Curiously, our attempts to apply this technique to the preparation of 2-substituted ethynylferrocenes were frustrated by our inability to form the dibromo addition

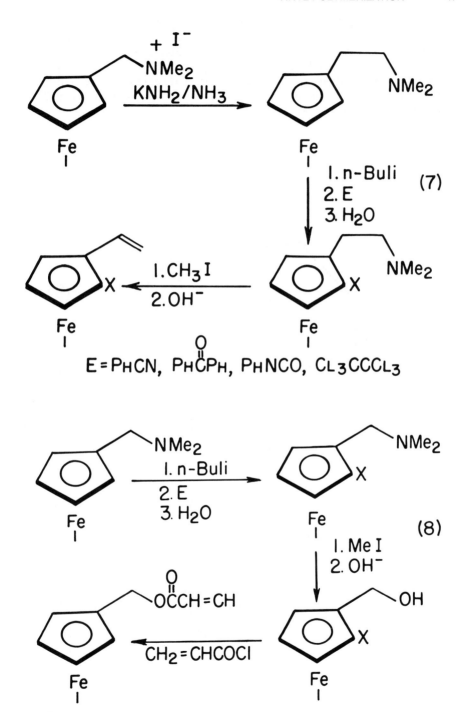

$$E = PhCN, \quad Ph\overset{O}{\overset{\|}{C}}Ph, \quad PhNCO, \quad CL_3CCCL_3$$

product of a suitable 2-substituted vinylferrocene (22). A more facile route which provided higher yields of ethynyl-ferrocene and related ethynylmetallocenes involves a modification of the procedure originally described by Arnold and Zemlica (24) and adapted to ethynylmetallocenes by Schlögl and Styrer (25), Rosenblum et al (26) and Rausch et al (27) (equation 9) (Table 4).

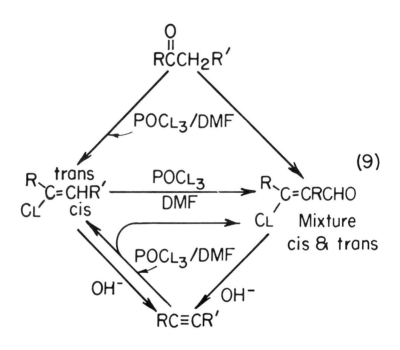

(9)

TABLE 4

Ethynylmetallocenes from
$$RCCH_2R' \text{ via Formylation}$$

R	R'	m.p.($^{\circ}$C)	% Yield
Fc	H	54–55	50
Fc	Ph	123–25	56
Fc	Fc	240–42	51
Rc	H	73–74	72

A high yield synthesis of diarylacetylenes described by Stephens and Castro (27) involves the coupling of a copper acetylide and an aryliodide in refluxing pyridine under nitrogen (equation 10). This procedure has been

$$RC{\equiv}CCu + R'I \xrightarrow[\text{reflux}]{\text{pyridine}} RC{\equiv}CR' \quad (10)$$

extended to the synthesis of a variety of ethynylmetallo-cenes. Table 5 lists the ethynylmetallocenes prepared by this method.

TABLE 5

Ferrocenyl Acetylenes Prepared
via the Copper Acetylide Route

R	R'	m.p.($^{\circ}$C)	% Yield
Fc	Ph	123–25	84
Fc	Fc	240–42	84
Fc	$2-C_4H_4S$	106–107	80
Rc	Fc	259–61	60
Fc	$1-C_{10}H_9$	162–62.5	83

Diarylacetylenes have also been prepared by dehydro-halogenation of dihalobibenzyls (28), oxidation of benzil dihydrazones with HgO, rearrangement of 1,1-diaryl-2-halo-ethylenes upon treatment with base (30) and by action of base on 5,5-diaryl-3-nitroso-2-oxazolidones (31).

In addition to their potential utility in polymeriza-tion processes, ethynylmetallocenes should also be useful in the synthesis of chlorinated olefins. For example ferrocenylacetylene reacts with HCl to produce a chloro-vinylferrocene in 20% yield (equation 11).

$$FcC{\equiv}CPh + HCl \longrightarrow \underset{Cl}{\overset{Fc}{\diagdown}}C{=}C\underset{Ph}{\overset{H}{\diagup}} \quad (11)$$

Speculative routes to a variety of novel but unknown ethynyl and vinyl derivatives of certain organometallic π-complexes can be put forward. A number of such complexes are potentially available through the syntheses developed in these laboratories. As illustrative examples, routes to vinylcyclooctatetraeneiron tricarbonyl and vinylnorbornadieneiron tricarbonyl can easily be devised based on routes developed during this study. Speculative preparations of certain polysubstituted organometallics can also be examined.

The keys to successful application and use of organometallic polymers are monomer cost, polymer properties and synthesis. Although many of the monomers described herein are presently expensive, large-scale production would decrease their price significantly. Since novel properties of a number of organometallic polymers have been identified, continued exploration of the synthesis of such organometallic complexes is to be desired.

REFERENCES

1. a. Arimoto, F.S., and Haven, A.C., Jr., J. Am. Chem. Soc. 77, 6295 (1955).
    b. Schlögl, K., and Mohar, A., Monatsh 92, 219 (1961).
    c. Fitzgerald, W.P., Jr., Diss. Abstr. 24, 2687 (1964).
2. cf. Pittman, C.U., Jr., in "Organometallic Reactions and Syntheses" (E.I. Becker and M. Tsutsui, Eds.) Vol. 6, p. 46. Plenum, New York, N.Y. 1977.
3. Hauser, C., Pruett, R., and Mashburn, A., J. Org. Chem. 26, 1800 (1961).
4. Coleman, L.E., Jr., and Rausch, M.D., J. Polymer Sci. 28, 207 (1958).
5. cf. ref. 2, pp. 47-48.
6. Rausch, M.D., and Siegel, A., J. Organomet. Chem. 11, 317 (1968).
7. a. Knox, G., and Pauson, P., J. Chem. Soc. 4610 (1961).
    b. Riemschneider, R., and Helm, D., Chem. Ber. 89, 155 (1956).
    c. Schlögl, K., and Fried, M., Monatsh. Chem. 95, 558 (1964).
8. cf. ref. 2, p. 46.
9. Pittman, C.U., Jr., Lai, J.C., Vanderpool, D.P., Good, M., and Prados, R., Macromolecules 3, 746 (1970).
10. Pittman, C.U., Jr., Voges, R.L., and Jones, W.R., Macromolecules 4, 291 (1971).
11. Pittman, C.U., Jr., Voges, R.L., and Jones, W.R., Macromolecules 4, 298 (1971).

12. Pittman, C. U., Jr., Voges, R. L., and Elder, J., _Macro-molecules_ 3, 302 (1971).
    a. Pittman, C. U., Jr., and Marlin, G. V., _J. Polymer Sci Chem. Ed._11, 2753 (1973).
13. cf. ref. 2, pp. 52-53.
14. cf. ref. 2, p. 54.
15. cf. ref. 2, p. 55.
16. a. Rausch, M.D., Moser, G.A., Zaiko, E.J., and Lipman, A.L., Jr., _J. Organometal. Chem._ 23, 185 (1970).
    b. cf. ref. 2, p. 51.
17. cf. ref. 2, p. 50.
18. Murdoch, H.D., and Weiss, E., _Helv. Chem. Acta_ 46, 1588 (1963).
19. a. Fujita, N., and Sonogashira, K., _J. Polymer Sci. Chem. Ed._ 12, 2845 (1974).
    b. cf. ref. 2, p. 57.
20. Chen, S.C., Lee, C.C., and Sutherland, R.G., _Syn. React. Inorg. Metal-Org. Chem._ (in press).
21. Horspool, W.M., and Sutherland, R.G., _Can. J. Chem._ 46, 3453 (1968).
22. Slocum, D.W., Jennings, C.A., Engelmann, T.R., Rockett, B.W., and Hauser, C.R., _J. Org. Chem._ 36, 377 (1971).
23. Slocum, D.W., and Sugarman, D., in Advances in Chemistry Series, No. 130, "Polyamine-chelated Alkali Metal Compounds" (Langer, A., Ed.) American Chemical Society, Washington, D.C., 1974, pp. 222-247.
24. Arnold, Z., and Zemlica, J., _Proc. Chem Soc._ 227 (1958).
25. Schlögl, K. and Steyrer, W., _Monatsh. Chem._ 94, 1520, (1965).
26. Rosenblum, M., Brawn, N., Papenmeier, J., and Applebaum, M., _J. Organometal. Chem._ 6, 173 (1966).
27. Rausch, M.D., Siegel, A., and Klemann, L.P., _J. Org. Chem._ 31, 2703 (1966).
28. Castro, C.E., and Stephens, R.D., _J. Org. Chem._ 28, 2163 (1963).
    Castro, C.E., and Stephens, R.D., _J. Org. Chem._ 28, 3313 (1963).
    Castro, C.E., Gaughn, E.J., and Owsley, D.C., _J. Org. Chem._ 31, 4071 (1966).
29. Smith, L.I., and Falkof, M.M., _Org. Synth._ 22, 50 (1942).
30. Cope, A.C., Smith, D.S., and Cotter, R.J., _Org. Synth._ 34, 42 (1954).
31. Coleman, G.H., Holst, W.H., and Maxwell, R.D., _J. Am. Chem. Soc._ 58, 2310 (1936).

THERMOMECHANICAL TRANSITIONS OF FERROCENE-CONTAINING POLYMERS

Y. Ozari* and J.E. Sheats**
Rider College

T.N. Williams, Jr.
Princeton University

C.U. Pittman, Jr.
University of Alabama

ABSTRACT

*Polyvinylferrocene (PVFc), random copolymers of vinyl-
ferrocene with styrene, methylacrylate and methyl methacrylate, and polyferrocenylmethyl methacrylate (PFMMA) and block
popolymers of ferrocenylmethyl methacrylate with styrene and
methyl methacrylate were examined between -180° and 250°C in
an inert atmosphere using fully automated torsional braid
apparatus. Thermal stability of the polymers was also examined thermogravimetrically. PVFc undergoes irreversible
changes in the temperature range 220° to 300°C nullifying
attempts to determine its glass transition temperature (Tg)
directly. The extrapolated value of the Tg of PVFc from Tg
values of its random copolymers was relatively low but the
extrapolated Tg of PFMMA from Tg values of its block copolymers was equal to that of PFMMA homopolymer. The Tg of PFMMA
is constant for $\overline{M}n$ ca. 10000 to ca. 150000. PFMMA is plasticized readily with m-bis (m-phenoxyphenoxy)benzene (PPB).
Dependence of Tg on composition of the plasticized system was
studied. In low concentrations (<18%) of plasticizer the
decrease of Tg is more drastic than in polystyrene.*

INTRODUCTION

Ferrocene containing polymers have been of special
interest for the past 20 years (1,2). Ferrocene possesses
great thermal stability, low toxicity and high extinction
coefficient for absorption of UV (3) and γ radiation (4). It
acts as an efficient quencher and as a sensitizer (5) in photochemical reactions. Mixed valance ferrocene-ferricinium
polymers formed by partial oxidation of ferrocene polymers

*Present address-GAF Corporation,1361 Alps Rd.,Wayne NJ 07470
**To whom correspondence should be addressed.

act as semiconductors (6,7). A number of authors have reported synthesis of oligomers and low molecular weight polymers containing ferrocenyl groups but little thermomechanical
data on these materials has been reported.

In general, introduction of ferrocenyl groups into
polymers results in a large increase in their glass transition temperatures, Tg (8-10), and a large increase in brit-
tleness of the materials that prevents their fabrication
without compounding. It is therefore desirable to characterize a series of ferrocene polymers of known composition and
molecular weight and to see quantitatively what effect the
ferrocenyl group has on thermomechanical properties.

Gillham (11,12) has perfected the technique of low
frequency (ca. 1 Hz) torsional braid analysis (TBA) as a method
for studying all thermal transitions of polymers in the
region -180°C to 350°C as well as obtaining accurate values
of the glass transition, Tg. The changes in tensile strength
as the polymer undergoes thermal transitions may be measured
quantitatively and irreversible changes during the transitions,
such as thermal decomposition or crosslinking, may be observed
if present. In general, a change in Tg or a change in the
shape of the curve (shear modulus versus temperature) during
repeated sweeps through the region of Tg is evidence of
irreversible change.

In this preliminary report thermomechanical data for
polyvinylferrocene, copolymers of vinylferrocene with styrene
and with methyl acrylate and a series of polyferrocenylmethyl
methacrylates (PFMMA) with a wide range of molecular weights
($\overline{Mn}$ = 2 x 10^3 to 250 x 10^3) are summarized.

The vinylferrocene polymers and copolymers were prepared by radical initiation, and are relatively low molecular
weight polymers with a wide molecular weight distribution.
The polyferrocenyl alkyl acrylates and their block copolymers
are a new class of anionically initiated polymers which have
been prepared recently (13.) High ($\overline{Mn}$ = 250 x 10^3) and relatively narrowly distributed molecular weights were achieved.

EXPERIMENTAL

The polymers were prepared as described in the literature (13-18). They were fractionated by precipitation with
petroleum ether from their benzene solutions (PVFc, PVFc/PS,
PVFc/PMA and PVFc/PMMA) or from their THF solutions (PFMMA,
PFMMA/PS and PFMMA/PMMA). $\overline{Mn}$ of PVFc was 12 x 10^3 ($\overline{Mw}/\overline{Mn}$ =
1.91), $\overline{Mn}$ of the copolymers was in the range 6000 to 15000
and Mn's of PFMMA and its block copolymers are listed in table
1. Blends of PFMMA with m-bis (m-phenoxy-phenoxy) benzene
(PPB) (Eastman Organic Chemicals) were obtained by dissolving
the polymer in benzene solutions of PPB. The plasticizer was
not removed from the blend even on heating to 200°C because
of its low volatility (BP = 273-276°C/1 torr).

TABLE 1

Molecular weights and compositions of PFMMA and its block co-polymers

Type	$\overline{Mn}$ (Total)	$\overline{Mw}/\overline{Mn}$ (Total)	Mole % FMMA	FMMA in Block $\overline{Mn}$	$\overline{Mw}/\overline{Mn}$
PFMMA	2810	1.54			
	8780	2.67			
	13100	2.80			
	21000	3.18			
	60000	2.63			
	144000	2.54			
	250000	2.63			
PFMMA/PS	9520	6.78	27.8	4800	1.87
PFMMA/PS	7150	5.39	48.8	4800	1.87
PFMMA/PS	14600	3.15	62.8	4800	1.87
PFMMA/PMMA	14700	2.22	33.3	7000	2.42
PFMMA/PMMA	20000	2.57	18.0	7760	2.22

TBA specimens were prepared by impregnating a two-inch multifilament glass braid with 10-20% solutions of polymers (g. polymer/ml. benzene). Solvents were removed from the specimen in the TBA apparatus in situ by heating to 140-200°C in He. Thermomechanical spectra were obtained using a fully automated torsional braid analysis system (11, 12, 19, 20). A hard-wired analog computer (data reducer) (19, 20) provides automatic alignments and initiation of oscillations of the pendulum, printing of the values of the temperature (mV), the period of the damped oscillations (P in sec) and logarithmic decrements, Δ [$\Delta = (1/n) \ln(Ai/Ai+n)$ where Ai is the amplitude of deformation of the ith oscillation and n is the number of oscillations between two fixed arbitrary boundary amplitudes in each wave]. The reduced data are also presented on an XY recorder versus the temperature (or time) as a logarithm of the relative rigidity $(1/p^2)$ [which is a good approximation (22) for the elastic shear modulus (G') and the logarithm of Δ].

TGA measurements were conducted on a duPont 950 Thermogravimetric Analyzer using 2-3 mg of polymer in N_2 atmosphere.

RESULTS AND DISCUSSION

Polyvinylferrocene and its random copolymers

The ferrocene group in the polymer causes an increase in Tg (8-10). An attempt has been made (10) to determine the Tg of PVFc (as 190°C) from broad curves of differential scanning calorimetry (DSC) experiments. TGA experiments indicate stability of the polymer, in weight, up to 410°C. The Tg determined by TBA (Fig. 1) seems to be 254°C.

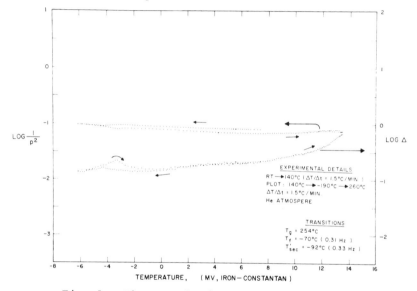

Fig. 1.   Thermomechanical spectra of PVFc

Repeat experiments in the temperature range ca 220° to 300° indicate irreversibility in the mechanical properties of the polymer. The Δ values of the damping peak decrease and the relative rigidity increases with time. This is a behavior typical of crosslinking which means that the observed peaks from DSC and TBA should not be referred only to the Tg of the material. It may arise from either Tg or crosslinking or both. In the spectra of PVFc there are no distinct secondary transitions (Tsec); however an apparent irreversible transformation (Tf) occurs at ca. -92°C (0.33 Hz) on cooling and at -70°C (0.31 Hz) on heating. This process appears also in the spectra of the copolymers of vinylferrocene (Figs. 2, 3, and 4). T_f diminishes when the experiment is recycled in the range -190°C to 0°C. It may be concluded that T_f is due to cracking of the brittle materials. As the temperature drops and it anneals as the temperature rises again.

The flat curves in the PVFc spectra reflect the restricted motion of the ferrocene side groups and of the backbone of the polymer due to the presence of a high density of large and bulky groups attached directly to its backbone. The material probably crosslinks or degrades at temperatures in the vicinity of its Tg. Transition from glassy to liquid states may facilitate these reactions.

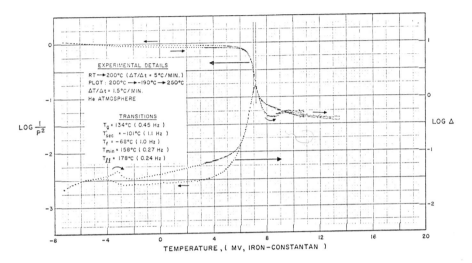

Fig. 2. *Thermomechanical spectra of PVFc/PS*

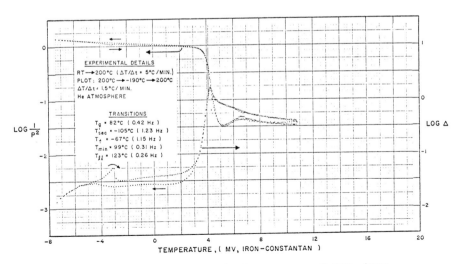

Fig. 3. *Thermomechanical spectra of PVFc/PMA*

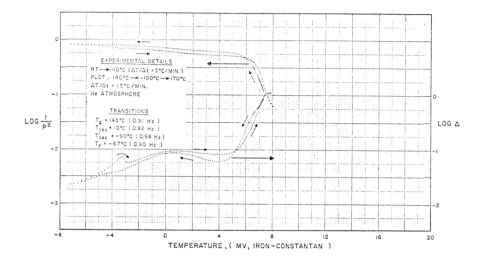

Fig. 4. Thermomechanical spectra of PVFc/PMMA

In order to evaluate the apparent Tg of PVFc, its random copolymers with styrene and with methyl acrylate of various compositions were examined (Fig. 5). The PVFc/PS copolymers are stable in weight up to 395°C to 405°C. In the spectra (Fig. 2), a sharp glass transition is observed in the range 114°C to 135°C (0.50 - 0.60 Hz). The transitions are lower than those expected for PVFc (ca. 250°) and higher than those of polystyrene (ca. 100°C) (21). Tmin occurs in PVFc/PS at 139°C to 158°C (0.24 - 0.31 Hz) and T ll occurs at 177°C to 195°C (0.19 -0.25 Hz) and no distinct secondary transition (Tsec) occurs in the temperature range -40°C to Tg. The spectra are similar to PS (21). The apparently characteristic T_f transition of the vinyl ferrocene at ca. -68°C to -80°C (1.0 - 1.4 Hz) appears in the spectra. A secondary transition Tsec' also appears at -101°C to -103°C (1.1-1.4 Hz) which decreases as the percentage of the vinyl ferrocene groups increases. According to this behavior of Tsec' and because of its appearance in the spectra of all the polymers examined in this investigation (Figs. 1, 2, 3, 4, 7 and 10)it is suggested that Tsec' reflects the motion of the ferrocene groups. When they are separated by other linkages their motion is facilitated and their Tsec' peak is relatively intense and narrow, but when they are crowded their motion is restricted and that broadens and decreases the intensity of their glass-glass transition peaks.

The random copolymers PVFc/PMA are stable in weight up to 355°C to 370°C. The pattern of their thermomechanical spectra is similar to that of PVFc/PS but the locations of Tg, Tmin

and Tsec transitions are different (Fig. 3). Tg appears in
the range 81°C to 100°C (0.42 - 0.65 Hz), Tll at 123°C to
149°C (0.23 -0.26 Hz), and Tmin at 98°C to 120°C (0.29 - 0.65
Hz). T_f and Tsec' are about the same as in PVFc/PS at -64°C
to -67°C (1.1 - 1.3 Hz) and at -99°C to 105°C (1.2 - 1.4 Hz)
respectively.

TGA measurements show that PVFc/PMMA decomposes at a relatively low temperature (155°C). Its thermomechanical spectrum
(Fig. 4) seems like an average of both homopolymers PVFc and
PMMA. A sharp Tg (145°C, 0.31 Hz) higher than that of PMMA
(22), Tsec (10°C, 0.92 Hz) as in PMMA (22) and T_f (-67°C, 0.90
Hz) and a broad Tsec' (-90°C, 0.98 Hz) due to the ferrocenic
part of the polymer.

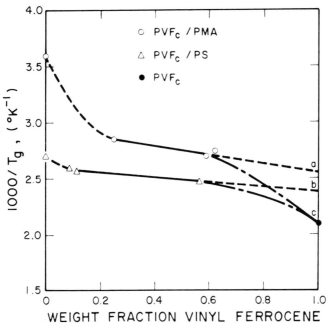

Fig. 5. Reciprocal of Tg vs. weight
fractions of vinyl ferrocene in PVFc/PS
and PVFc/PMA

A plot of reciprocal of Tg vs. weight fractions of vinylferrocene in the copolymers PVFc/PS and PVFc/PMA (Fig. 5) indicates, as expected, an increase in Tg as the number of ferrocene groups increases, but the extrapolated Tg values (points
b and a in Fig. 5, 145°C and 120°C, respectively) are lower
than the Tg measured for PVFc (point C in Fig. 5, ca. 250°C).
The reactivity ratios for vinylferrocene are considerably lower
than for other monomers (17) so that ferrocene units must appear
as single groups or very short chains in the structure of the
copolymer. Therefore, the extrapolated low Tg values obtained

should be attributed to Tg for oligomers rather than for high polymers of vinylferrocene. A large increase of Tg in molecular weight range 2000 to 10000 has been shown in the case of polystyrene (21) Fig. 6) and of PFMMA and also is expected for PVFc.

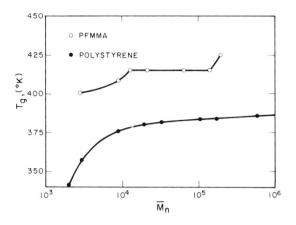

Fig. 6. Tg (^{O}K) vs. Logarithm of $\overline{M}$n of PFMMA and of polystyrene (ref. 21).

Polyferrocenylmethyl methacrylate and its block copolymers:

PFMMA is stable in weight up to 225°C to 250°C. In this polymer the ferrocenyl group is not bound directly to the backbone as in PVFc; therefore, as expected, the movement of the polymeric chain is less restricted and in the spectrum (Fig. 7) of the polymer a sharp Tg (129°C to 151°C, 0.44 Hz) dependent on molecular weight is observed. This Tg is higher than that of PMMA (22) and lower than that expected for PVFc. Two secondary transitions are observed (for Mn = 8780): at 5°C(0.89 Hz) and at -110°C (0.90 Hz). The former is similar to Tsec of PMMA (isotactic) (22) and the latter coincides with Tsec' observed for PVFc/PS, PVFc/PMA and PVFc/PMMA. The apparent transition T_f of PVFc and its copolymers does not appear in the spectra of PFMMA. At extreme low or high molecular weights of PFMMA, Tg is affected by the molecular weight but in a large region of $\overline{M}$n (13,100-144,000), Tg does not vary (Fig. 6). It is suggested that the polymers are interlocked (11). The large and bulky ferrocenylmethyl acrylate side groups, as compared with phenyl groups of polystyrene (21) are more effective in intermolecular interlocking.

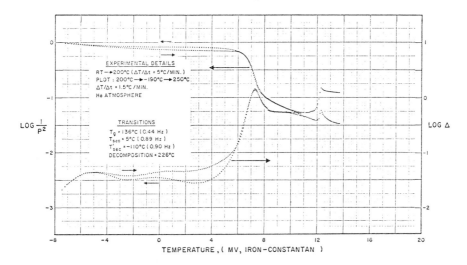

Fig. 7. Thermomechanical spectra of PFMMA.

TGA measurements indicate stability of the block copolymers PFMMA/PS and PFMMA/PMMA up to ca. 250°C. In the low molar
percentage of FMMA in the block (27.8% and 48.8%) there is
evidence for two Tg peaks in the thermomechanical spectra
(112°C, 127°C, 0.53 Hz, 0.35 Hz and 106°C, 129°C, 0.37 Hz,
0.30 Hz respectively) and for higher percentage (62.8%) only
one sharp Tg peak appears (133°C, 0.38 Hz, Fig. 8). This
behavior may be due to limited compatibility of the two parts
of the chain (PFMMA and PS) in the block. At low per centages
of PS there is full compatibility. In both blocks PFMMA/PS
and PFMMA/PMMA the temperature of the major peak increases as
the molar percentage of PFMMA increases. The extrapolated Tg
value for 100% PFMMA (Fig. 9) (140° to 153°C) coincides with
the observed values for PFMMA homopolymer. This supports our
assumption that Tg of a homopolymer can be extrapolated from
its copolymers only if enough segments of the same kind appear
in the structure of the copolymer.

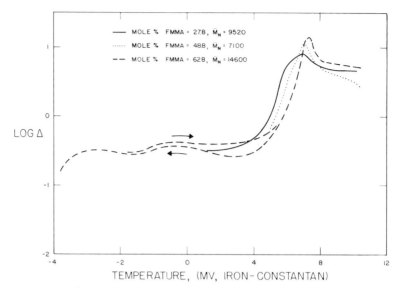

Fig. 8.   *Logarithmic decrement vs. tempera-*
ture of PFMMA/PS block copolymers

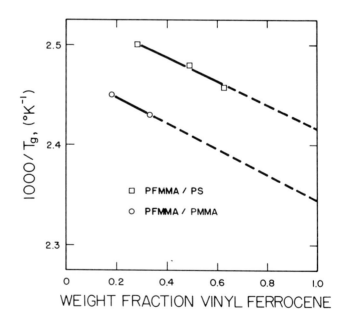

Fig. 9.   *Reciprocal of Tg vs. weight fraction*
of vinyl ferrocene in PFMMA/PS and
PFMMA/PMMA block copolymers.

Benzene solutions of PFMMA with PPB are clear and no phase separation can be observed when films are cast from these solutions. A single Tg peak appears in the thermomechanical spectra of the PFMMA-PPB blends (Fig. 10) which is between Tg's of PFMMA and PPB. It is evident, therefore, that PPB acts as a plasticizer for PFMMA. The polymer-plasticizer interactions affect also the secondary transitions (Fig. 10). Tsec is eliminated by the presence of PPB and Tsec' is shifted (-165°C, 1.05 Hz).

In the low concentration region of PPB (< 18%) Tg of PFMMA-PPB is more affected by the presence of the plasticizer than polystyrene-PPB (Fig. 11)   i.e. 10% of PPB causes a decrease in Tg of polystyrene by ca. 23°C and of PFMMA by ca. 46°C.

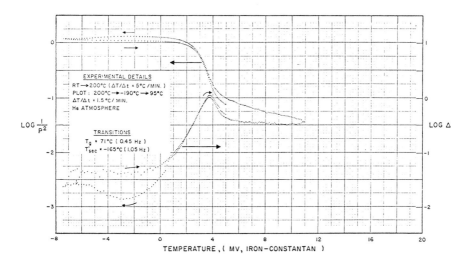

Fig. 10.   Thermomechanical spectra of PFMMA/plasticizer blend (80:20 by weight, respectively)

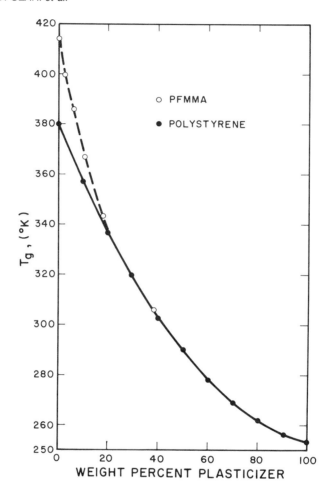

Fig. 11.   *Tg (ᵒK) vs. weight (%) of plasticizer
in PFMMA/plasticizer blends and in
polystyrene/plasticizer blends (ref. 21).*

ACKNOWLEDGEMENT

Financial support was provided by Miles Laboratories, Elkhart, Indiana.

REFERENCES

1. Neuse, E.W., in "Advances in Macromolecular Chemistry", Vol. 1, M.W. Paska, Ed., Academic Press, New York, N.Y. 1968.
2. Pittman, C.U., Jr., J. Paint Technology, 39, 585 (1967).
3. Schmitt, R.G. and Hirt, R.C., American Cyanamid Co., Air Force WADC Technical Reports 59-354, 60-704 and 61-298.
4. McIlhenny, R.C. And Honigstein, S.A., July 1965 Air Force Report No. AFMLTR-65-294, AD476623, Contract No. AF-33-615-1694.
5. Dannesburg, J.J. and Richards, J.H., J. Amer. Chem. Soc., 88,4781 (1966).
6. Cowan, D.O., Park, J., Pittman,C.U., Jr., Sasaki, Y., Mukherjee, T.K. and Diamond, N.A., J. Amer. Chem. Soc., 94, 5110 (1972).
7. Pittman, C.U., Jr. and Saski, Y., Chem. Lett. 383 (1975).
8. Pittman, C.U., Jr. Lai, J.C., Vanderpool, D.P., Good, M., and Praclos, R., Macromolecules, 3, 746 (1970).
9. Tinker, A.J., Barrie, J.A., and George, M.H.,Polymer Letters, 13 (1975).
10. Sasaki, Y., Walker, L.L., Hurst, E.L.,and Pittman,C.U.,Jr. J. Polymer Sci., 11, 1213 (1973).
11. Gillham, J.K., CRC Critical Reviews in Macromolecular Science, 1, 83 (1972).
12. Gillham, J.K., A.I.Ch.E. Journal, 20 (6), 1066 (1974).
13. Pittman, C.U., Jr, and Hirao, A., J. Polym. Sci., in press.
14. Arimoto, F.S. and Haven, A.G., J. Am. Chem. Soc., 77, 6295 (1955).
15. Baldwin, M.G. and Johnson, K.E., J. Polymer Sci., (A-1) 5, 2091 (1967).
16. Pittman, C.U., Jr. J. Paint Technology, 43, 29 (1971).
17. Pittman, C.U., Jr., J. Polymer Sci., B6, 19 (1968).
18. Pittman, C.U., Jr., unpublished results.
19. Bell, C.L.M., Gillham, J.K. and Benci, J.A., Am. Chem. Soc. Poly. Preprints, 15 (1), 542 (1974).
20. Bell, C.L.M., Gillham, J.K., and Benci, J.A., Soc. Plastics Engineers, Annual Tech. Conf., San Francisco, Cal., SPE Tech. Papers, (20), 598 (1974).
21. Gillham, J.K., Benci, J.A. and Boyer, R.F., Polymer Engineering and Science, 16 (5), 357 (1976)
22. Stadnicki, S.J., Gillham, J.K. and Hazony,Y., J.Appl.Poly. Sci. (1977) in press.

THE PREPARATION, POLYMERIZATION AND COPOLYMERIZATION
OF SUBSTITUTED DERIVATIVES OF CYNICHRODENE
$(\eta^5\text{-}C_5H_5)Cr(CO)_2NO$

Charles U. Pittman, Jr. and Thane D. Rounsefell
Department of Chemistry, University of Alabama

John E. Sheats and Bruce H. Edwards
Department of Chemistry, Rider College

Marvin D. Rausch and Eric A. Mintz
Department of Chemistry, University of Massachusetts

ABSTRACT. The novel vinyl organometallic monomer (η^5-vinyl-cyclopentadienyl)dicarbonylnitrosylchromium (hereafter called vinylcynichrodene)*, was prepared and homopolymerized and copolymerized in solution and neat using azo initiators. Reactivity ratios were obtained in radical initiated copolymerizations with styrene and N-vinyl-2-pyrrolidone. When M_2 was styrene, $r_1 = 0.30$ and $r_2 = 0.82$. The value of the Alfrey-Price parameter e for vinylcynichrodene was -1.98 and the resonance interaction parameter Q was 3.1. Copolymerizations with N-vinyl-2-pyrrolidone (M_2) gave $r_1 = 5.3$, $r_2 = 0.079$ and e = -2.07. Thus, vinylcynichrodene is an exceptionally electron-rich vinyl monomer resembling vinylferrocene and vinylcymantrene. Vinylcynichrodene readily copolymerized with vinyl-cymantrene.

I. INTRODUCTION.

The synthesis of organometallic polymers (1-5) and studies of their novel properties (1-7) have accelerated in scope in recent years. The radical-initiated addition homopolymerization and copolymerization of a variety of transition-metal-containing monomers has been under intense study in our laboratory. For example, studies of organometallic monomers including η^6-styrenetricarbonylchromium (8), vinylcymantrene (1-9), vinylferrocene (10), η^6-(benzyl acrylate)tricarbonylchromium (11), ferrocenylmethyl acrylate (12,13), η^6-(2-phenylethyl

---

*The trivial name <u>cynichrodene</u> is proposed by analogy to the shortened names cymantrene and benchrotrene for η^5-cyclopentadienyltricarbonylmanganese and η^6-benzenetricarbonylchromium, respectively.

acrylate)-tricarbonylchromium (14), as well as 2-ferrocenyl-
ethyl acrylate (15) and methacrylate (15) have now appeared.
Recently, a comprehensive review of vinyl addition polymerization of organometallic monomers appeared (16).

Addition polymerization studies of organometallic carbonyl monomers are rare, indeed, compared with metallocenes
(3). Organochromium monomer syntheses and polymerizations are
restricted to the three examples: η^6-styrenetricarbonylchro-
mium (8), η^6-(benzyl acrylate)tricarbonylchromium (11) and
η^6-(2-phenylethyl acrylate)tricarbonylchromium (14). Vinylcymantrene (1) *1*. is isoelectric with vinylcynichrodene,
2. Since vinylcymantrene's copolymerization reactivity with
a series of organic monomers has now been defined (1,17), it
was of interest to see how changing the metal fragment from
Mn(CO)$_3$ to Cr(CO)$_2$NO would affect the monomer reactivity.

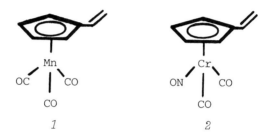

1   *2*

We now report the addition of homopolymerization and Copolymerization of the novel monomer, vinylcynichrodene, *2*. To
our knowledge, this represents the first example of the polymerization of an organometallic monomer containing a nitrosyl
(NO) substituent. Furthermore, we demonstrate that changing
the metallic fragment from Mn(CO)$_3$ to Cr(CO)$_2$NO did not exert
drastic effects on the reactivity of the vinyl group. In both
1 and *2* the vinyl group was exceedingly electron rich.

II. RESULTS AND DISCUSSION.

It has been found that transition metal organometallic
functions, when attached to a vinyl group, often exert enormous effects on addition polymerization reactivity (16,17).
For example, vinylcymantrene *1* and vinylferrocene, when class-
ified according to the Alfrey-Price $\underline{Q}$-$\underline{e}$ scheme (18,19), exhibit $\underline{e}$-values of -1.99 and -2.1, respectively, in copolymerizations with styrene. This means the vinyl group in both
these organometallic monomers is more electron-rich than that
of 1,1-bis-($\underline{p}$-anisyl)ethylene ($\underline{e}$ = -1.96) (19). In order to
extend the series of organometallic monomers, vinylcynichro-
dene, *2*, was synthesized according to Scheme I.

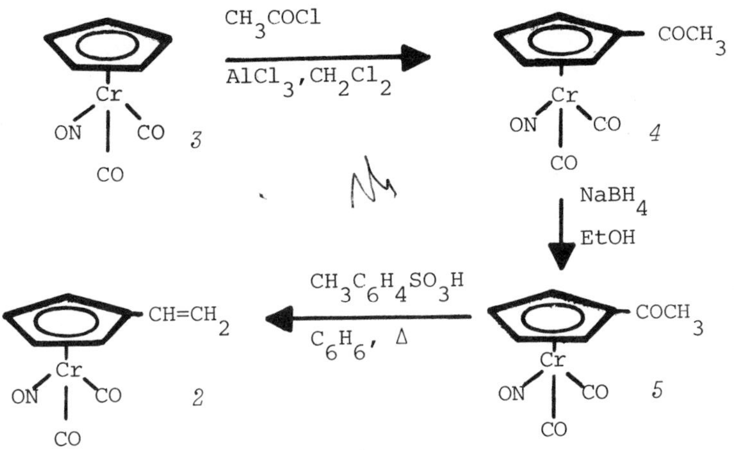

$$CH_3COCl \xrightarrow{AlCl_3, CH_2Cl_2}$$

Cr
ON | CO *3*
CO

Cr
ON | CO *4*
CO   COCH$_3$

NaBH$_4$
EtOH

Cr
ON | CO *2*
CO   CH=CH$_2$

$$\xleftarrow[C_6H_6, \Delta]{CH_3C_6H_4SO_3H}$$

Cr
ON | CO *5*
CO   COCH$_3$

Scheme 1

Cynichrodene, *3*, (Scheme I) was first prepared by Fischer et al in 1955 (20). Since that time very little chemistry of this complex has appeared. It was acylated in good yield by Fischer and Plesske (21) but the chemistry remained largely unexplored. Cynichrodene was acetylated in 50-65% yield, followed by sodium borohydride reduction of the carbonyl group (88%) and p-toluene-sulfonic acid-catalyzed dehydration (70-86%) of the resulting alcohol in the presence of hydroquinone as a radical inhibitor. Vinylcynichrodene is a red liquid, bp 79-80°/0.3 mm Hg, which was conveniently purified by filtration through alumina and repeated vacuum distillation.

Purification is particularly important because alcohol, *5*, readily forms its ethyl ether if traces of ethanol are present in the dehydration step. Furthermore, small amounts of the dimeric ether of *5* are also formed during dehydration. The ir and nmr spectra of *2*, used in this work, were in accord with structure and a satisfactory analysis was obtained. In addition, glc studies of *2* were made to monitor its purity.

Cynichrodenylmethyl acrylate, *8*, another useful organochromium monomer, is prepared as shown in Scheme II. Cynichrodene, *3*, is converted to the thioester *6* under Friedel-Crafts conditions by means of methyl chlorothiolformate and aluminum chloride in dry methylene chloride. Reduction of *6* with sodium borohydride in refluxing methanol gives cynichrodenylcarbinol, *7*. Reaction of *7* with two equivalents of pyridine produces cynichrodenylmethyl acrylate, *8*.

Vinylcynichrodene was homopolymerized and copolymerized with styrene, N-vinyl-2-pyrrolidone, and vinylcymantrene (Scheme 3). Azo initiators were used in each case. A series of copolymerizations with styrene were carried out to determine the reactivity ratios with *2*. Styrene was chosen because it

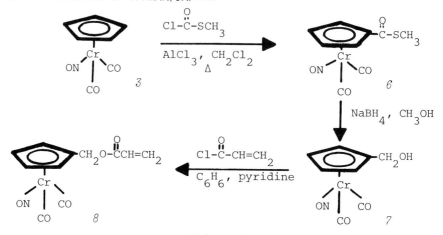

Scheme 2

had previously been used in reactivity ratio studies with vinylcymantrene, *1* (1,17). Furthermore, styrene is a moderately
electron rich monomer (e = -.80). We anticipated that if *2*
were a very electron rich monomer resembling *1* the use of
electron attracting comonomers (i.e., acrylonitrile or methyl
acrylate) might result in failure of the Q-e scheme, similar to
that observed for both vinylferrocene and *1* when copolymerized
with electron attracting monomers (17).

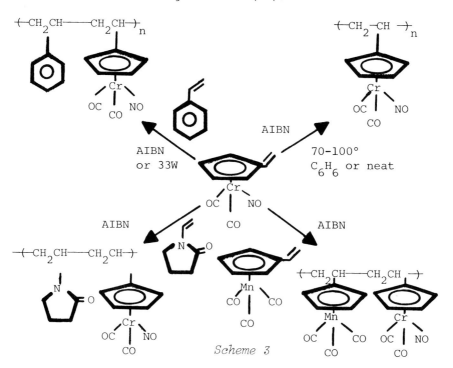

Scheme 3

Only small amounts of 2 were available. Thus, the disappearance of monomers was followed, quantitatively, by glc
using internal standards and electronic integration techniques.
In this way, the incorporation of monomers into the copolymer
could be followed as a function of percent conversion. Several
composition-conversion data points could be obtained for each
individual experiment, allowing large numbers of points to be
achieved with minimum expenditure of monomer. The composition-
conversion data obtained when styrene was M_2 is summarized on
Figure 1. This data was fitted to the <u>integrated</u> form of the
copolymer equation by the <u>nonlinear least squares</u> method of
Tidwell and Mortimer (23,24) using programs we prepared for
use on a Univac 1110 computer (25). In addition to using the
nonlinear fitting technique, $M_1°:M_2°$ ratios were applied in
the vicinity of the <u>optimum values</u> for the reactivity ratios
obtained. The importance of this choice of experiments has
been discussed by Tidwell and Mortimer (23,24).

From the data in Figure 1, the value of $r_1 = 0.30$
(0.287 - 0.313) and $r_2 = 0.82$ (0.774 - 0.875) (where $M_2 =$
styrene). Using the relationship $r_1r_2 = \exp. -(\underline{e}_1 - \underline{e}_2)^2$ (18,
19) the Price $\underline{e}$-value for vinylcynichrodene is -1.98. This

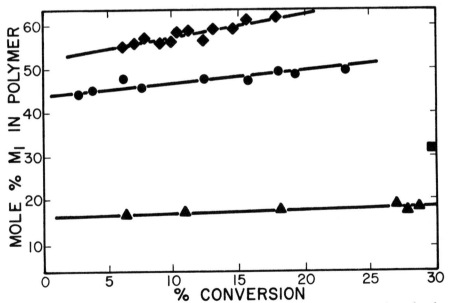

*Fig. 1. Composition-conversion plots for example vinyl-
cynichrodene (M_1) styrene copolymerizations.*
▲ $M_1°:M_2° = 71.3:28.7$
■ $M_1°:M_2° = 48.4:51.6$
● $M_1°:M_2° = 69.9:30.1$
◆ $M_1°:M_2° = 83.1:16.9$

may be compared to $\underline{e}$-values of -2.1 and -1.99 for vinylferrocene and vinylcymantrene, respectively in copolymerizations
with styrene. Clearly, vinylcynichrodene is exceptionally
electron rich and its reactivity closely resembles that of
vinylcymantrene.

In order to appreciate just what this high negative value
of $\underline{e}$ means, it is instructive to consider the $\underline{e}$-values for
selected monomers: (19)   maleic anhydride, +2.25; acryloni-
trile, +1.20; styrene, -0.80; p-N,N-dimethylaminostyrene,
-1.37; and 1,1-bis-(p-anisyl)ethylene, -1.96. Quite obviously, the electron-rich vinylcynichrodene (-1.98) would be expected to undergo cationic polymerization while resisting
anionic polymerization. Indeed, reaction with BuLi or LiAlH$_4$
failed to give polymer supporting this conclusion.

The value of the resonance parameter $\underline{Q}$ for vinylcynichrodene, from these styrene copolymerizations, was 3.13.
This may be compared to the monomers shown below (19). Clearly,

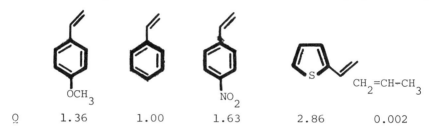

| $\underline{Q}$ | 1.36 | 1.00 | 1.63 | 2.86 | 0.002 |

a strong resonance interaction of the vinyl group with the
cyclopentadienyl ring is indicated. It is particularly noteworthy to recognize the similarity in the values of $\underline{Q}$ for
2-vinylthiophene (2.86) and vinylcynichrodene (3.13). In
both, the vinyl group is conjugated to an aromatic five-mem-
bered ring. However, the electron-donating ability of the
$(\eta^5\text{-}C_5H_4)Cr(CO)_2NO$ substituent ($\underline{e}$ = -1.98) is substantially
greater than that of the 2-thiophenyl group ($\underline{e}$ = 0.80).

Reactivity studies were also performed with N-vinyl-2-
pyrrolidone (M_2). This monomer was chosen because it has a
somewhat more electron rich vinyl group ($\underline{e}$ = -1.14) than
styrene ($\underline{e}$ = -.80) and a much smaller value of $\underline{Q}$ (0.14 versus
1.00 for styrene). It was of interest to check if $\underline{Q}$-$\underline{e}$ studies
based on copolymerizations with N-vinyl-2-pyrrolidone would
give an $\underline{e}$-value for vinylcynichrodene close to that obtained
in the styrene copolymerizations. Composition-conversion
studies, obtained in the same manner as in the styrene copolymerizations, are listed in Table 1. Using the nonlinear least
squares technique, the reactivity ratios were r_1 = 5.34 (5.08
- 5.60) and r_2 = 0.079 (0.073 - 0.084) and $\underline{e}_1$ = -2.07. The
agreement between the values of $\underline{e}$ in copolymerizations of 2
with styrene and N-vinyl-2-pyrrolidone (i.e., -1.98 and -2.07)

TABLE 1

Composition-Conversion Data for Copolymerizations of
Vinylcynichrodene, 2, with N-Vinyl-2-pyrrolidone (M_2) *

Run	Monomer 2 in Feed Mole %	Conversion %	Monomer 2 in Copolymer Mole %
1a	5.44	2.56	32.7
1b	5.44	10.82	36.8
1c	5.44	13.32	28.6
1d	5.44	13.72	30.1
1e	5.44	13.14	30.0
1f	5.44	14.02	28.0
1g	5.44	17.22	23.8
1h	5.44	20.65	24.5
1i	5.44	43.32	10.2
2a	22.36	3.13	64.8
2b	22.36	3.12	68.6
2c	22.36	6.41	69.0
2d	22.36	6.85	69.6
2e	22.36	7.23	65.7
2f	22.36	7.18	68.6
2g	22.36	7.55	67.3
2h	22.36	10.92	62.5
2i	22.36	11.12	66.5
2j	22.36	11.48	65.9
3a	3.33	9.79	21.9
3b	3.33	10.36	20.8
3c	3.33	10.12	20.8
3d	3.33	10.11	21.2
3e	3.33	9.88	21.3
3f	3.33	10.70	19.4
3g	3.33	10.91	18.5
4a	27.87	2.14	68.0
4b	27.87	5.66	67.0
4c	27.87	6.04	69.6
4d	27.87	6.40	69.5
4e	27.87	6.39	71.1
4f	27.87	6.55	69.9
4g	27.87	6.77	68.9
4h	27.87	6.62	68.7
4i	27.87	7.09	70.0

*From this data the calculated reactivity ratios are
r_1 = 5.34 and r_2 = 0.079.

is remarkable and strengthens the assignment of e at about -2.
    Vinylcymantrene was copolymerized with vinylcynichrodene
but only a single $M_1°:M_2°$ ratio was employed (Table 2). Thus,

the reactivity ratios were not calculated.

TABLE 2

Composition-Conversion Data for Copolymerizations of Vinylcynichrodene, 2, with Vinylcymantrene

Monomer 2 in Feed Mole %	Conversion %	Monomer 2 in Copolymer Mole %
30.84	7.5	34.5
30.84	9.3	35.3
30.84	9.9	35.4
30.84	10.4	35.0
30.84	10.8	35.1
30.84	10.5	34.7
30.84	10.6	35.3
30.84	17.0	34.6
30.84	20.1	34.5
30.84	21.8	35.7
30.84	21.5	34.8
30.84	21.5	35.1
30.84	21.5	34.9

Radical initiated homopolymerizations of vinylcynichrodene were slow. Only a 6.4% yield of polymer was isolated after 49h at 70° in the ethylacetate using AIBN. Similarly, only an 11% yield was isolated when neat vinylcymantrene was heated to 100° (with AIBN) for 2h followed by further initiator addition and another 3h at 100°. Solution homopolymerizations give low molecular weights (below 35,000) and molecular weights increased when the monomer concentration increased or the initiator concentration decreased.

Additional copolymerization studies are currently underway involving 2 and other monomers with a view toward obtaining polymers with novel properties. In this regard, cynichrodene, *3*, and several of its derivatives have been shown to effectively catalyze reactions such as hydrogenation (26). Further, based on recent studies by Alt and Herberhold (27,28), the possibility of preparing chiral polymers is clearly available.

III.   EXPERIMENTAL SECTION.

1-Hydroxyethylcynichrodene (*5*).

Acetylcynichrodene (2.2 g) (15) was dissolved in 25 ml

of ethanol, 0.168 g of sodium borohydride was added, and the mixture was stirred for ca. 45 min. After this period, 10-20 ml of 6N sodium hydroxide solution was added, the solvent removed, and the product extracted with ethyl ether. The ether was removed, leaving a red oil; bp 105-110°/0.3 mm Hg; 1.95 g (88%).

Vinylcynichrodene (2).

1-Hydroxyethylcynichrodene (0.5 g), 5 mg of hydroquinone, and 0.05 g of p-toluenesulfonic acid were dissolved in 50 ml of benzene, and the mixture refluxed for 1.25 h. The solvent was removed and the resulting oil was extracted with hexane and subsequently filtered through silica gel, eluting several times with hexane and then 5:1 hexane-ether. The solvent was removed, leaving a red oil; bp 79-80°/0.3 mm Hg; 0.41 g (86%). Its IR spectrum gave strong ν_{CO} bands at 1952 and 2030 cm^{-1} and a ν_{NO} band at 1700 cm^{-1}. Its nmr was in accord with structure and its analysis found: C, 47.41; H, 3.10; Cr, 22.46; N, 6.36%. Calculated for $C_9H_7CrNO_3$: C, 47.17; H, 3.08; Cr, 22.69; N, 6.11.

Homopolymerization.

Vinylcynichrodene (1.072 g, 4.6 mmol), AIBN (0.0024 g, 0.15 mmol) and ethyl acetate (1.45 g) were weighed into a vial which was degassed by two freeze-thaw-pump cycles. The vial was sealed and placed in a constant temperature bath at 70° for 49h. The ethyl acetate solution was then added dropwise to a large excess (50 ml) of pentane to precipitate the polymer. The precipitated homopolymer was filtered and dried in vacuo, to give 0.069 g (6.4% yield) of polyvinylcynichrodene. The IR of the polymer contains carbonyl stretching bands at 2020 and 1930 cm^{-1} and a nitrosyl stretch at 1680 cm^{-1}. In another run, neat vinylcynichrodene (0.5 g) was mixed with AIBN (5 mg) followed by heating to 100° for 2h followed by the addition of another 5 mg of AIBN. After an additional 3h, methanol (2 ml) was added, the precipitated polymer was washed with excess methanol and dried, giving 56 mg (11%) yield of polyvinylcynichrodene. Gel permeation chromatography of this sample revealed a broad molecular weight distribution with a low $\overline{M}_n$. (See Figure 2).

Copolymerizations.

All copolymerizations were performed in dried, deoxygenated benzene at 70° in a 3 ml Minivert ® sample vials stirred magnetically. In each case, the disappearance of the monomers was followed by glc. A typical run is described for one vinyl-cynichrodene/styrene copolymerization. Vinylcynichro-

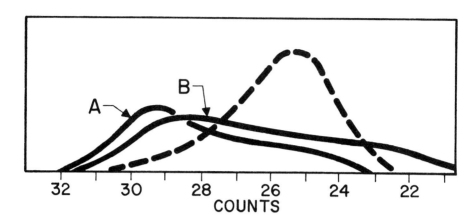

*Fig. 2.   Gel permeation chromatograms of vinylcynichro-
dene polymers.*

        *A.   Polyvinylcynichrodene. Bulk polymerization
at 100°, with AIBN initiator.*

        *B.   Polyvinylcynichrodene. Polymerized in ethyl
acetate to 6.4% conversion using AIBN initia-
tor.*

-----   *Copolymer of vinylcynichrodene and N-vinylpyrrolidone (No. 1i, Table 1).*

dene, styrene, naphthalene (used as an internal glc standard),
and benzene were weighed into a 3 ml vial. A small aliquot
was removed into a one ml vial. Then both vials were degassed
under nitrogen by five freeze-flush-thaw cycles. The initiator, 33W (i.e., 2.2'-azobis(2,4-dimethyl-4-methoxyvaleroni-
trile)), was added by syringe (in benzene solution) several
times during the polymerization. At least one hour was allowed after each initiator injection before sampling the reaction for glc analysis. In some reactions, AIBN was used as
the initiator and in these cases it was added prior to degas-
sing. The same procedure was used when the comonomer was N-
vinylpyrrolidone or vinylcymantrene.

Analysis of Copolymerization Reactions.

A typical example is described for a styrene copolymerization. The analysis of monomer remaining in the reaction as
a function of time was carried out by glc (1/8" x 50 cm, OV-
101 (5%) on Chromasorb G-HP, 100/120 mesh, flow rate 22 cc.
min.$^{-1}$, injection 80°, detector 160°, filament 170°, tempera-
ture-programmed 1 min. at 45°, 5° min.$^{-1}$ to 110°, 6 min. at
110°). Several injections were made into the column prior to
gathering data. A Varian Associates Model 3700 gas chromato-
graph was used and electronic integration was performed using

a Hewlett-Packard Model 3380A recorder-integrator.

Samples from the 1 ml reference vial were alternately injected with samples from the polymerizing solution, which were withdrawn from the 3 ml vial being held at constant temperature in a water bath. Three injections were averaged to determine the amount of each monomer remaining as a function of time. The internal standards used included quinoline (for N-vinylpyrrolidone copolymerizations) and napthalene (for both styrene and vinylcymantrene copolymerizations). A sample gel permeation chromatogram is shown in Figure 2.

IV.   ACKNOWLEDGMENT.

Acknowledgment is made to the Donors of the Petroleum Research Fund, administered by the American Chemical Society (MDR), to the National Science Foundation (MDR) and to the Office of Naval Research (CUP) for support of this research.

V.   REFERENCES.

1.   Pittman, Jr., C.U., Marlin, G.V., and Rounsefell, T.D., Macromolecules, 6, 1 (1973).
2.   Pittman, Jr., C.U., Chem. Tech., 1, 417 (1974).
3.   Pittman, Jr., C.U., J. Paint Technol., 43, 29 (1971).
4.   Neuse, E. W. and Rosenberg, H., J. Macromol. Sci. Rev., 4, 1 (1970).
5.   Neuse, E.W. and Rosenberg, H., "Metallocene Polymers," Chapter 2. Marcel Dekker, New York, 1970.
6.   Cowan, D.O., Park, J., Pittman, Jr., C.U., Sasaki, Y., Mukherjee, T.K., and Diamond, N.A., J. Am. Chem. Soc., 94, 5110 (1972).
7.   Bilow, N., Landis, A.L., and Rosenberg, H., J. Polym. Sci., A-1, 7, 2719 (1969).
8.   Pittman, Jr., C.U., Grube, P.L., Ayers, O.E., McManus, S.P., Rausch, M.D., and Moser, G.A., J. Polym. Sci., A-1, 10, 379 (1972).
9.   Pittman, Jr., C.U., and Grube, P.L., J. Polym. Sci., A-1, 9, 3175 (1971).
10. Lai, J.C., Rounsefell, T.D., and Pittman, Jr., C.U. J. Polym. Sci., A-1, 9, 651 (1971).
11. Pittman, Jr., C.U., Voges, R.L., and Elder, J., Macromolecules, 4, 302 (1971).
12. Lai, J.C., Rounsefell, T.D., and Pittman, Jr., C.U., Macromolecules, 4, 155 (1971).
13. Ayers, O.E., McManus, S.P., and Pittman, Jr., C.U., J. Polym. Sci., A-1, 11, 1201 (1973).
14. Pittman, Jr., C.U., and Marlin, G.V., J. Polym. Sci., A-1, 11, 2753 (1973).

15. Pittman, Jr., C.U., Voges, R.L., and Jones, W.R., Macromolecules, 4, 291, 298 (1971).

16. Pittman, Jr., C.U., "Vinyl Polymerization of Organic Monomers Containing Transition Metals", a chapter in Organometallic Reactions, Vol. 6, p. 1. (E. Becker and M. Tsutsui, Eds.), Plenum Press, New York, 1977.

17. Pittman, Jr., C.U., and Rounsefell, T.D., Macromolecules, 9, 936 (1976).

18. Price, C.C., J. Polym. Sci., 3, 772 (1948).

19. Brandrup, J., and Immergut, E.H., "Polymer Handbook," Interscience, New York, 1966.

20. Fischer, E.O., Beckert, O., Hafner, W., and Stahl, H.O., Z. Naturforsch., 10B, 598 (1955).

21. Fischer, E.O., and Plesske, K., Chem. Ber., 94, 93 (1961).

22. Mintz, E. A., Rausch, M.D., Edwards, B.H., Sheats, J.E., Rounsefell, T.D., and Pittman, Jr., C.U., J. Organomet. Chem., 0000 (1977).

23. Tidwell, P.W., and Mortimer, G.A., J. Macromol. Sci. Rev. Macromol. Chem., 5, 135 (1970).

24. Tidwell, P.W., and Mortimer, G.A., J. Polym. Sci., A-1, 3, 369 (1965).

25. Pittman, Jr., C.U., and Rounsefell, T.D., in "Computers in Chemistry and Instrumentation" (J.S. Mattson, H. C. MacDonald, Jr., and H.B. Mark, Eds.), Vol. 6, pp. 145-197. Marcel Dekker, New York, 1977.

26. Mintz, E., and Rausch, M.D., unpublished results.

27. Herberhold, M., and Alt, H., J. Organomet. Chem., 42, 407 (1972).

28. Herberhold, M., Alt. H., and Kreiter, C.G., J. Organometal. Chem., 42, 413 (1972).

ORGANOMETALLIC CONDENSATION POLYMERS

Charles E. Carraher, Jr.
Wright State University

I. INTRODUCTION

Siloxanes are the only organometallic polymers utilized widely industrially. The potential of organometallic polymers is seen by considering these siloxanes, which offer a positive combination of electrical, chemical, and mechanical properties not common to any other class of (organic) polymers. They exhibit high oxidative and thermal stabilities, low power losses, unique rheological properties, high dielectric strengths, and are relatively inert to most ionic reagents.

This review concerns the synthesis and characterization of organometallic polymers utilizing condensation processes that can often be considered as extensions of inorganic coordination chelations, as noted below. It will emphasize work done by our group.

Preparative conditions are generally not unusual but are often narrower than for nonorganomatallic systems. Low temperature condensation techniques, namely, interfacial and solution techniques, are normally utilized in an attempt to avoid undesirable thermally induced competing reactions.

II. POTENTIAL

Reasons for desiring inclusion of metal atoms and metal-containing moieties are many and varied. For instance, compounds containing the Cp_2M moiety (where M = Fe, Ti, Zr, Hf) are reported to be stabilizers to UV radiation; thus inclusion into paint mixtures, etc., might impart to the outdoor paint better wearability regarding UV photo stability (for instance, 1, 2). Most compounds containing tin are antifungal and antibacterial and would impart to rugs, paint, clothing, etc., possessing tin-containing polymers, mildew and rot resistance (for instance, 3-8).

Most of the metal-containing monomers utilized by us possess catalytic properties. Thus specific metal-containing polymers might offer not only specific catalytic activity but also certain stereoregulating properties resulting from inherent steric requirements present in the metal-containing moiety.

Both ferrocene and cobalticinium moieties act as haptens when bound to proteins. Thus polymers (modified natural or synthetic) containing these moieties might offer useful biological activities (9). Compounds containing tin or arsenic coupled with sugars, purines, pyrimidines, etc. might also offer useful (controlled release) biological activities. For instance, product I was found to be active against only very select bacteria in a mode consistant with active metabolism of the base portion, releasing the toxic arsenic-containing moiety which "kills" the bacteria (10).

I.

A number of titanium-containing polymers synthesized by the interfacial technique exhibit a characteristic known as "anomolous fiber formation." Certain of the fibers offer good weight retention (80% to 1200°C) with some dimensional stability. They also possess internal "hooks" (11-14). Exceptional strength may be expected.

The use of a polymerization system for the analytical removal of materials is also possible. We have removed the uranyl ion to ca 10^{-6}M utilizing dioximes and diacids which can form polyoximes and polyesters (15-17).

Other uses include as photosensitive media in photo-copying devices, adhesives, flame retarders, electrically conductive and semiconductive applications and applications requiring thermal stability.

The above is not intended to be exhaustive, but only offers some of the possible diverse uses of condensation-type organometallic polymers.

III.   EXTENT

In theory a quite wide range of metal-containing moieties can be employed, dictated by desired end properties. In actuality, severe limitations are evident. First, synthesis of many desired monomers is difficult or not yet accomplished. For instance a good synthesis of tungstanyl chloride, WO_2Cl_2 is yet to be found. Synthesis of a number of desirable "sandwich" nickel, iron, cobalt compounds has yet to be achieved. Synthesis of certain arsenic and antimony compounds has proven dangerous, with explosions occurring.

Second, the physical and chemical properties, including actual structure, of many potentially useful monomers are unknown, little know, or conflicting.

Third, selection of reaction conditions is critical and often quite limited. For instance, synthesis of antimony polyoximes at concentrations normally employed in interfacial systems takes ca nine hours at a stirring rate of ca 10,000 rpm; whereas product is rapidly formed (ca within 30 secs) em-ploying much greater concentrations of reactants. Formation

of zirconocene poly(cobalticinium dicarboxylate) decreases
with stirring time after 30 seconds with no product obtained
if stirring is allowed to proceed past several minutes.
Synthesis of Group IV B polyethers has thus far been accomplished only with aromatic diols with aqueous systems.
Use of excess Group IV B Cp_2MX_2 monomers results in the
formation of crosslined polyamidoximes rather than the
(generally desired)linear products. Synthesis of antimony
polyesters has occurred only with certain antimony compounds,
being dependent on such factors as inherent reactivity,
aqueous solubility and hydrolysis rate.

Fourth, the inherent reactivity of metal-containing
reactants must be considered. Thus far, fully positively
charged and highly electron deficient metal sites (generally
halo derivatives) have proven most successful in the synthesis
of such organometallic products.

Many Lewis Bases have been successfully employed. These
include the following:

$$\underset{\underset{R}{}}{\overset{HON}{C}} - R - \underset{\underset{R}{}}{\overset{NOH}{C}} \qquad \underset{\underset{H_2N}{}}{\overset{HON}{C}} - R - \underset{\underset{NH_2}{}}{\overset{NOH}{C}} \qquad HS - R - SH$$

oximes   amidoximes   thiols

$$HO - R - OH \qquad \underset{\underset{R}{}}{H_2N - NH} \qquad H_2NNH - \overset{X}{\overset{\|}{C}} - R - \overset{X}{\overset{\|}{C}} - NHNH_2$$

diols   hydrazines   hydrazides and thiohydrazides

$$H_2N - R - NH_2 \qquad H - N\underset{R}{\overset{R}{\Big(}} N - H \qquad \overset{\ominus}{O_2C} - R - \overset{\ominus}{CO_2}$$

(primary)amines   (secondary)amines   acid salts

$$H_2N - \overset{X}{\overset{\|}{C}} - NH_2$$

ureas and thioureas

A number of metal-containing Lewis Bases have also been
employed. These include mainly the iron and cobalt containing metallocenes. Such as the following:

$$HO-R_2C-CpFeCp-CR_2-OH$$
$$\underset{R}{\overset{HO-N}{}}C-CpFeCp-C\underset{R}{\overset{N-OH}{}}$$

$$HO_2C-CpFeCp-CO_2H$$
$$HO_2C-CpCoCp-CO_2H \quad X^{\ominus}$$

Polymers successfully modified include

$+CH_2-CH_2-NH+$ (21) $+CH_2-CH+$ via $\longrightarrow$ $+CH_2-CH+$ (7,22,23,26)

$\begin{matrix} | \\ C{\equiv}N \end{matrix}$

$\begin{matrix} | \\ C \\ H_2N \nearrow \diagdown NOH \end{matrix}$

$+CH_2-CH+$ (25,27-29) $+CH_2-CH+$ (30) $+CH_2-CH+$ (8,24)

$\begin{matrix} | \\ OH \end{matrix}$ $\begin{matrix} | \\ C=O \\ | \\ Cl \end{matrix}$ $\begin{matrix} | \\ C=O \\ | \\ O \ominus \end{matrix}$

Metal-containing Lewis Acids successfully employed include

Cp_2HfCl_2 (11,18-20,22,28,40,43)
Cp_2ZrCl_2 (11,19,20,22,28,35,37,39,43,46,49)
Cp_2TiCl_2 (11-13,19,22,24,28,36,41,43,44,47,48)
UO_2^{+2} (15-17)
R_3SbX_2 (18,31-33)
R_3AsX_2 (10,19,31,34)
R_3BiX_2 (18,31)
R_3PbX_2 (50)
R_2SnX_2 (4-8,19-21,29,30,42,45)
R_2GeX_2 (3,4,7,8,19,20)
R_2SiX_2 (3,4,7,8,19,20)

IV. GENERAL PHYSICAL PROPERTIES

Relative to organic polymers, organometallic polymers are
soluble in fewer solvents and to lesser extents within these
solvents (for instance 18-20,51). Many are insoluble in all
solvents attempted. These poor solubilities are probably due
to a combination of related factors including a.high cohesive
(secondary bonding) forces between chains, b. highly crystalline nature of most products, and c. a peculiar combination
of bonding offered by the polymers - a combination of both
nonpolar and low to high polar contributions. Dipolar aprotic
solvents as DMSO, DMF, acetone, HMPA, TEP and Sulfolane-W
have proven to be the best solvents. For the three families
studied, range and extent of solubility appears to decrease
with increase in the size of the metal (solubility C>Si>Ge>
Sn>Pb; Ti>Zr>Hf [tentative] and P>As>Sb>Bi [tentative]).
Most products can be obtained as powders. Films can be
cast from many of the prodcuts containing Si and Ge. Films
can be cast from a number of otherproducts provided the
product is prepared in the presence of a plasticizer. Organic
polymers modified utilizing organomettic monomers are generally rubbery with the formation of tough films possible

from linear products.

The products are hydrophobic and resistant to hydrolysis until "wet" by addition of acetone, etc.

Thermal degradation generally occurs via a series of kinetically and/or thermodynamically controlled plateaus characteristic of differences in the thermal stability of bonds present in the products. Melting (Tg and/or Tm) may or may not occur prior to initial degradation, depending on the product. Degradation occrring via oxidative modes in air is typical.

Products (tested) containing tin have all shown at least some antibacterial or antifungal activity.

Many products exhibit bulk resistivites within the range of 10^4 to 10^{10} ohm-cm and thus can be considered for semi-conductor applications.

V.   CONCLUSION

The inclusion of organometallic and inorganic monomers in into polymers has begun. Potential users are emerging with most current useage limited to the siloxanes. The use of many as yet unused metal-containing reactants is possible within limits matching such factors as monomer properties and preparation, reaction conditions and desired properties. Because of current cost and scarcity, initial users will be limited to "low bulk amount" applications, speciality items and to use as catalysts.

VI. REFERENCES

1.   E. Neuse and H. Rosenberg, "Metallocene. Polymers," Marcel Dekker, N.Y., 1970.
2.   C. Pittman, J. Paint Technol., 39, 585 (1967).
3.   C. Carraher and R. Dammeler, Makromolekulare Chemie, 135, 197 (1970) and 141, 251 (1971).
4.   C. Carraher and R. Dammeier, J. Polymer Sci., 8, 3367 (1970) and 10, 413 (1972).
5.   C. Carraher and D. Winter, Makromolekulare Chemie, 141, 237 (1971) and 141, 259 (1971) and 152, 55 (1972).
6.   C. Carraher and G. Scherubel, J. Polymer Sci., 9, 983 (1971).
7.   C. Carraher and L. Wang, Makromolekulare Chemie, 152, 43 (1972).
8.   C. Carraher and J. Piersma, J. Appl. Polymer Sci., 16, 1851 (1972).
9.   T. Gill and L. Mann, J. Immunology, 96, 906 (1966).
10. C. Carraher, W. Moon, T. Langworthy, C. Gausch and W. Mayberry, unpublished results.
11. C. Carraher, Chem. Tech. 741 (1972).
12. C. Carraher and P. Lessek, Eup. Polymer J., 8, 1339 (1972).

13. C. Carraher and R. Nordin, Makromolekulare Chemie, 164, 87 (1973).
14. C. Carraher and S. Bajah, Polymer 14, 42 (1973).
15. C. Carraher and J. Schroeder, Polymer Letters, 13, 215 (1975).
16. C. Carraher and J. Schroeder, Polymer Preprints, 16, 659 (1975).
17. C. Carraher and P. Fullenkamp, unpublished results.
18. F. Millich and C. Carraher (Eds.) "Interfacial Synthesis," Marcel Dekker, N.Y., 1976.
19. C. Carraher, "Reactions on Polymers," J. Moore, Ed., Plenum, N.Y., 1974.
20. C. Carraher, Inorganic Macromolecules Reviews, 1, 271 and 287 (1972).
21. C. Carraher and M. Feddersen, unpublished work.
22. C. Carraher and L. Wang, Angew, Makromolekular Chemie, 25, 121 (1972).
23. C. Carraher and L. Wang, J. Macromol. Sci., A7, 513 (1973).
24. C. Carraher and J. Piersma, Makromolekulare Chemie, 153, 49 (1972).
25. C. Carraher and L. Torre, Angew, Makromolekulare Chemie, 21, 207 (1972).
26. C. Carraher and L. Wang, J. Polymer Sci., A-1, 9,. 2893 (1971).
27. C. Carraher and L. Torre, J. Polymer Sci., A-1, 9, 975 (1971).
28. C. Carraher and J. Persma, J. Macromol. Sci., A7, 913 (1973).
29. C. Carraher and J. Persma, Angew, Makromolekulare Chemie, 28, 153 (1973).
30. C. Carraher and M. Moran, unpublished work.
31. C. Carraher and H. Blaxall, unpublished work.
32. C. Carraher and H. Blaxall, Polymer Preprints, 16, 837 (1975).
33. C. Carraher, J. Sheets and H. Blaxall, Polymer Preprints 16, 655 (1975).
34. C. Carraher and L. Hedlund, Polymer Preprints, 16, 847 (1975).
35. C. Carraher and J. Reimer, Polymer, 13, 153 (1972).
36. C. Carraher and R. Nordin, J. Polymer Sci., A-1, 10, 521 (1972).
37. C. Carraher and J. Reimer, J. Polymer Sci., A-1, 10, 3367 (1972).
38. C. Carraher and G. Scherubel, Makromolekulare Chemie, 160, 259 (1972).
39. C. Carraher, Eup. Polymer J., 8, 215 (1972).
40. C. Carraher, Angew. Makromolekulare Chemie, 28, 145 (1973).

41. C. Carraher and J. Sheats, Makromolekulare Chemie, 166, 23 (1973).
42. C. Carraher and D. Winter, J. Macromolecular Sci., A-7, 1349 (1973).
43. C. Carraher, Makromolekulare Chemie, 166, 31 (1973).
44. C. Carraher and J. Lee, J. Macromol. Sci. Chem., A9, 191 (1975).
45. C. Carraher, G. Peterson, J. Sheats and T. Kirsch, Makromolekular Chemie, 175, 3089 (1974).
46. C. Carraher and L. Jambaya, J. Macromolecular Sci. Chem., A8, 1249 (1974).
47. C. Carraher and R. Frary, Br. Polymer J., 6, 255 (1974).
48. C. Carraher and S. Bajah, Polymer, 15, 9 (1974).
49. C. Carraher and R. Nordin, J. Applied Polymer Sci., 18, 53 (1974).
50. C. Carraher and C. Deremo, Organic Coatings and Plastics Chemistry, in press.
51. C. Carraher, Polymer Preprints, 17, 365 (1976).
52. C. Carraher, unpublished results.

I gratefully acknowledge support through American Chemical Society - Petroleum Research Foundation Grant number 9126-B1,3-C.

SYNTHESIS AND PROPERTIES OF POLYMERS CONTAINING COBALTICINIUM-1,1'-DICARBOXYLIC ACID

John E. Sheats
Rider College

Cobalticinium salts, isoelectronic with the corresponding ferrocene derivatives, possess great thermal stability and marked resistance to degradation by strong acids or bases or by oxidation (1). In contrast to the extensive research on ferrocene polymers (2), much less work has been done on cobalticinium polymers. The possibility for synthesis of a wide range of polymers containing cobalticinium salts was proposed and the properties of cobalticinium salts were discussed extensively at the previous Symposium on Organometallic Polymers in 1971 (3). This paper will summarize the experiments which have been performed in the six years since the previous paper was presented.

I.   POLYESTERS OF COBALTICINIUM 1,1'-DICARBOXYLIC ACID.

Cobalticinium 1,1'dicarboxylic acid, I, is readily prepared by permanganate oxidation of 1,1'-dimethylcobalticinium salts and can be converted readily to the acid chloride II and to the ester III.   The counterion is hexafluorosphosphate, PF_6^-, which is inert and also imparts solubility for II and III in both water and polar aprotic solvents.   When solutions of II in acetonitrile were added (4) to solutions of 1,4 butanediol and to solutions of 1,4-bis(hydroxymethyl) benzene also in acetonitrile, low molecular weight polymers $IV_{a,b}$ were formed.

I X=OH IV_a R=$(CH_2)_4$

II X=Cl   b   R=p-CH_2-C_6H_4-CH_2-

III X=OCH$_3$

A portion of the polymeric material was soluble in acetone and DMF. The peak in the I.R. at 1770 cm^{-1} characteristic of II disappeared and a new peak for the ester linkage appeared at 1725 cm^{-1}. Intrinsic viscosities of 0.034 - 0.078 dl/g for samples of IVa and 0.095 - 9.120 dl/g for IVb were obtained, which correspond to molecular weights Mn of 2500-4000. Gel permeation chromatography confirmed these results and indicated the presence of some material with molecular weight up to 8000. Thus the material is oligomeric with DP$_n$ = 4-20. Exchange of the PF$_6^-$ ion by Cl$^-$ during the reaction is extensive, and in some cases almost complete.

In a further attempt to obtain higher molecular weight samples of IVa and IVb (5), the dimethyl ester, III, was heated with the corresponding alcohols in a sealed tube at 175-200° for 2 hr. Polymeric material was obtained, part of which was soluble in DMF and part insoluble. The DMF soluble material had $\overline{Mn}$ = 2700 , $\overline{Mw}$ = 4500. No exchange of the PF$_6^-$ anion occurred. Analysis of the insoluble fraction showed it to be high in carbon and low in cobalt, indicative of extensive decomposition.

Attempts to prepare polyamides by pyrolysis of diamine salts of I *in vacuo* at 175-200° led to decomposition or recovery of starting material (5). Thus, attempts to prepare and characterize high molecular weight polyesters and polyamides of I have so far proven unsuccessful. The materials, if obtained, are likely to be insoluble in most common solvents and therefore will be difficult to characterize.

II. PREPARATION OF ORGANOMETALLIC COPOLYMERS OF I.

Carraher et al (6-10) have prepared a series of copolymers of organometallic monomers R$_2$MCl$_2$ with dicarboxylic acids. Notable among them are the titanocene (10) polymers which readily form long filaments. Nearly all of these polymers are insoluble in water and most organic solvents and are therefore intractable. A series of copolymers have been prepared with I in the hope that,by adding charges along the chain, the solubility of the polymer in polar solvents would be increased.

The following comonomers have been used: R_2SnCl_2 (11,12),
where R = phenyl, t-butyl, vinyl, n-octyl, ethyl and methyl,
$(C_6H_5)_2$ $TiCl_2$ $(C_6H_5)_2$ Ti $(PF_6)_2$ (13), $(C_6H_5)_2$ $ZrCl_2$ (14) and
$(C_6H_5)_3$ $SbCl_2$ (15). Compound I is dissolved in aqueous base
and stirred rapidly under nitrogen with a solution of the comonomer in $CHCl_3$ at 10,000-25,000 rpm for 1-2 min. Yields of
30-80% have been obtained.

The goal of achieving greater solubility has, unfortunately, not been achieved. The polymers containing I are insoluble in water and have limited solubility in most organic
solvents. 2-Chloroethanol has proven to be the best solvent
but solubilities are still less than 0.5%. Intrinsic viscosity
measurements at high dilution in 2-chloroethanol indicate
molecular weights of approximately 8×10^4 for the Ti (13) and
Zr (14) polymers and $2-5\times10^3$ for the Sn (11,12) and Sb (15)
polymers. The Sn polymers prepared from uncharged dicarboxylic acids had previously been shown to be oligomers with molecular weights in the range $2-5\times10^3$ (16).

The driving force for polymerization appears to be formation of an insoluble hydrophobic precipitate. The PF_6^- ion
plays a key role in rendering the polymer insoluble. When
PF_6^- is replaced by Br^-, Cl^- or NO_3^-, poor yields of polymers
are obtained with Sn or Sb monomers (12). If I is dissolved
in aqueous base and used immediately, polymer yields are good
and little exchange of PF_6^- occurs, but if the solution of I
is allowed to sit for 30 min. or more, ion exchange occurs.
Apparently I dissolves as a tight ion pair with the PF_6^- ion
and dissociation occurs slowly. Since no coordination site is
available on the cobalt atoms,the PF_6 ion must remain in a secondary coordination sphere. This unique behavior of the PF_6^-
ion may explain why other salts of I do not produce good
yields of polymers.

The resonance, inductive and steric effects of the R
groups also influence the polymer yield. The yields of Sn
polymers vary in the following order (11) C_6H_5 > vinyl >
t-butyl > n-octyl $\approx$ ethyl > methyl. The order of substituents
corresponds well with σ^+ values for the substituents, being
greatest for the most positive values of σ^+. The yield also
correlates reasonably well with the steric bulk of the substituents. Since the Sn and Sb monomers are hydrolytically
unstable, the larger substituents would shield the ester
linkages from hydrolysis long enough for the material to
polymerize sufficiently to become insoluble. The halide
salts of I have a greater solubility, so that the polymers
remain in a solution long enough for the Sn and Sb monomers
to be hydrolyzed to unreactive products. The greater stabil-

ity of the Ti and Zr monomers toward hydrolysis may facilitate
formation of polymers with higher molecular weight. All of
the polymers, however, may be degraded back to I by prolonged
stirring with aqueous salt solutions (11-15).

Thus the goal of obtaining soluble, high molecular
weight polymers containing I has not been achieved. A recent
communication (17) has described a method of preparing soluble photocrosslinkable polymers from organometallic compounds
by the procedure shown below. Procedures of this type may
also be applicable for polymers containing I.

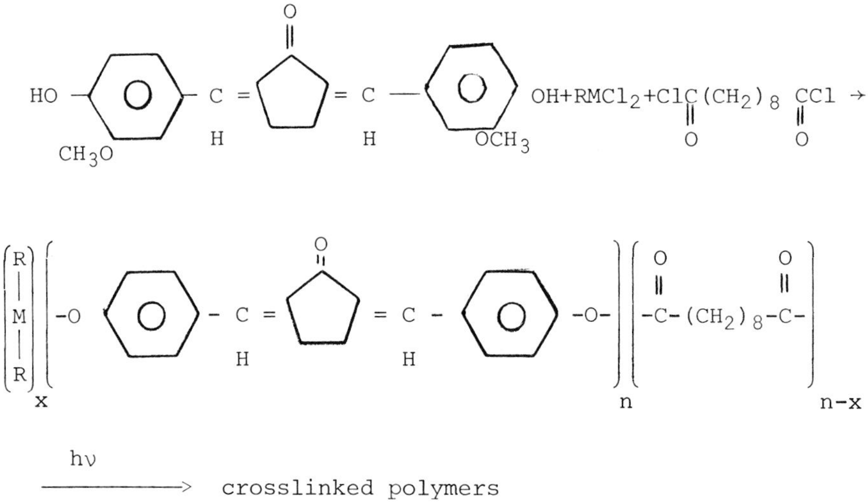

$$\xrightarrow{h\nu} \text{crosslinked polymers}$$

III.   IR SPECTRA OF POLYMERS.

The IR spectrum of I contains a sharp peak at 3130 cm^{-1}
for the C-H stretch, a broad band at 3300-2500 cm^{-1} for the
O-H stretch, a sharp C=O at 1730 cm^{-1}, C-O at 1270 cm^{-1}, other
peaks at 1475, 1380, 1340, 1165 and 1030 cm^{-1} and an intense
peak at 850 cm^{-1} for the vibrations of the PF_6^- ion. Upon formation of the Na salt, the C=O peak shifts to 1610 cm^{-1}. The
polymers contain a broad absorption in the region around 3400
characteristic of M-OH end groups, a carbonyl peak at 1550-
1610, a cyclopentadienyl peak around 1480 and the PF_6^- peak at
850. The extent of exchange of the PF_6^- anion by Cl^- or other
anions can be estimated by the ratio of the areas of the peaks
at 1480 and 850. Since the polymers are insoluble in most
common solvents UV and NMR spectra can not be obtained.

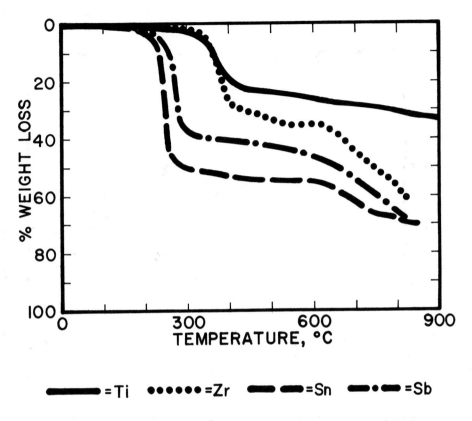

FIGURE 1

TGA ANALYSIS OF COPOLYMERS OF $(C_5H_4CO_2^-)_2Co^+PF_6$
WITH $(C_5H_5)_2TiCl_2$, $(C_5H_5)_2 ZrCl_2$, $(C_4H_9)_2 SnCl_2$
AND $(C_6H_5)_3SbCl_2$ UNDER N_2. THE SAMPLES WERE
HEATED AT 10°C/MIN.

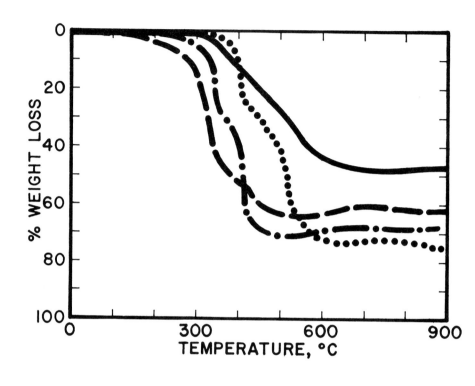

FIGURE 2

TGA ANALYSIS OF COPOLYMERS OF $(C_5H_4CO_2^-)_2Co^+PF_6$
WITH $(C_5H_5)_2TiCl_2$, $(C_5H_5)_2 ZrCl_2$, $(C_4H_9)_2 SnCl_2$
AND $(C_6H_5)_3SbCl_2$ IN AIR. THE SAMPLES WERE
HEATED AT 10°C/MIN.

IV.   TGA AND DSC ANALYSIS OF POLYMERS.

TGA analyses of the Ti, Zr, Sn and Sb polymers are shown
in Fig. 1-2. The Ti and Zr polymers (13,14) show initial
degradation at 200-225° leading to approximately 15-20% weight
loss in air of N_2 at 300°. Under N_2 the Ti polymer shows
little further weight loss up to 900°C, while the Zr polymer
shows 60% weight loss at 900°. In air rapid weight loss begins at 450-500° and leads to 40% weight loss for Ti and 70%
weight loss for Zr above 600° with no further change up to
900°C. Conversion of the polymer to Co_2O_3 and ZrO_2 with loss
of all other materials would correspond to 70% weight loss.
Thus, a significant fraction of the organic materials remain
in the case of Ti at 900° while the Zr polymers may well be
degraded completely to inorganic oxides. The Sb polymers (15)
begin degradation at 200° in both air and N_2, leading to 30%
weight loss at 325° in N_2 and a 58% weight loss in air. Under
N_2, degradation continues slowly but 43% of the material still
remains at 800°. The Sn polymers are much less stable, showing initial degradation at 200° in air and in N_2 and 50%
weight loss above 300°. Mass spectrographic analyses of the
products given off during pyrolysis and analyses of the residues remaining may be undertaken to clarify the nature of the
processes taking place.

The DSC analyses of all four polymers show endothermic
transitions taking place under N_2 and strongly exothermic
transitions taking place in air for Ti at 350°, 420° and 600°,
Zr at 470° and $(C_6H_5)_2$ Sn at 320° and 420° and the Sb polymers
at 310°, 375° and 460°. Mass spectrometric analysis would
also be useful in determining what is happening.

V.   REFERENCES

1. Sheats, J. E. and Rausch, M.D.,J. Org. Chem.   35, 3245
    (1970).
2. Neuse, E. and Rosenberg, H., "Metallocene Polymers",
    Marcel Dekker, New York (1970).
3. Sheats, J. E., Amer. Chem. Soc. Org. Coatings and Plastics
    Chem.   31(2), 277 (1970).
4. Pittman, Jr., C. U., Ayers, O. E., McManus, S. P., Sheats,
    J. E. and Whitten, C. E., Macromolecules 4,360 (1971).
5. Pittman, Jr., C. U., Ayers, O. E., Suryanarayanan, B.,
    McManus, S. P., and Sheats, J. E., Makromol. Chem. 175,
    1427 (1974).
6. Carraher, C., and Schrubel, G., J. Polymer Sci. A-1, 9
    983 (1971).
7. Carraher, C. and Winter, D., Makromol. Chem. 141, 259
    (1971).

8. Carraher, C., Macromolecules 4, 263 (1971).

9. Carraher, C., and Dammeier, R., Makromol. Chem. 141, 245 (1971).

10. Carraher, C., J. Polymer Sci. A-1, 9, 3661 (1971).

11. Carraher, Jr., C. E., Peterson, G. F., Sheats, J. E., and Kirsch, T., Makromol. Chem. 175, 3089 (1974).

12. Carraher, Jr., C. E., Peterson, G. F., Sheats, J. E., and Kirsch, T., J. Macromol. Sci.-Chem. A8(6), 1009 (1974).

13. Carraher, Jr., C. E., and Sheats, J. E., Makromol. Chem. 166, 23 (1973).

14. Sheats, J. E., Carraher, Jr., C. E., Bruyer, D., and Cole, M., Amer. Chem. Soc. Org. Coatings and Plastics Chem. 34(2) (1974).

15. Sheats, J. E., Carraher, Jr., C. E., and Blaxall, H. S., A.C.S. Polymer Preprints 16(1), 655 (1975).

16. Carraher, Jr., C. E., Inorg. Macromol. Revs. 1, 271 (1972).

17. Borden, D. G., Am. Chem. Soc. Org. Coatings and Plastics Chem. 37(1), 72 (1977).

POLY(COBALTICENIUM CARBOXAMIDES)[*]

Eberhard W. Neuse
University of the Witwatersrand
Johannesburg, South Africa

The polycondensation of 1,1'-dicarboxycobalticenium chloride with aromatic diamines in molten antimony trichloride at 150-175°C gives polyamides containing the cobalticenium tetrachloroantimonate salt system in the recurring unit. The products are isolated initially as the mixed hexafluorophosphate-tetrachloroantimonate salts, which, upon further treatment with NH_4PF_6, can be obtained as the pure hexafluorophosphates. IR and NMR data confirm the linear polyamide structures 1 . The polymers are soluble and film-forming; viscometric and vapor pressure-osmometric investigations on these polyelectrolytes are hampered, however, by polydissociation in both neutral and acidic solvents.

I. INTRODUCTION.

Polyamides containing the cobalticenium cation in the recurring unit were prepared previously by Pittman *et al.* (1) using a melt polymerization technique; yet much decomposition occurred in these reactions, and well-defined soluble polyamides were not obtained. In an effort to improve both the composition and the solubility properties of cobalticenium polyamides (and, later on, of other cobalticenium-containing condensation polymers), we investigated the use of molten antimony trichloride as a polymerization medium for the reaction of 1,1'-dicarboxycobalticenium salts with aromatic diamines (equation 1). The results of this study are reported below.

(eq. 1)

1a: Ar = $\underline{p}$-C$_6$H$_4$

1b: Ar = $\underline{p}$-C$_6$H$_4$-$\underline{p}$-C$_6$H$_4$

1c: Ar = $\underline{p}$-C$_6$H$_4$-CH$_2$-$\underline{p}$-C$_6$H$_4$

---

[*]Metallocene Polymers, XXXIV.

II.   RESULTS AND DISCUSSION.

Lewis acids of the chalcogen halide type, such as arsenic and antimony chlorides or fluorides, proposed earlier both as polymer solvents (2) and as polymerization media in the preparation of polypyrrolones (3), were recently used in this laboratory for the synthesis of non-polymeric cobalticenium carboxamides (4). Antimony trichloride proved especially efficacious in these cobalticenium condensation reactions. In contrast to polyphosphoric acid, or even the less aggressive ethylated polyphosphoric acid, both of which were found in other investigations to cause cobalt-ring bond cleavage at the reaction temperatures required, the antimony halide, forming homogeneous solutions of cobalticenium compounds at temperatures as low as 80°C, proved indifferent in this respect, although insignificant elimination of substituents was observed in that work (4) after extended heating at 170°C. In the present study, the reactants of equation 1 (Ar = 1,4-phenylene, 4,4'-biphenylene, 4,4'-methylenediphenylene) were heated for 20-45 hr at 150-175°C in antimony trichloride solution (ca. 20% solute concentration). Three typical condensation experiments are summarized in Table 1. The products ($\underset{\sim}{1}$, $A^- = SbCl_4^-$) were worked up under conditions essentially as elaborated for the non-polymeric reactions (4) and were purified by reprecipitation from aqueous or aqueous dimethyl sulfoxide (DMSO) phase in the presence of PF_6^- anion; the yields given in the table refer to the reprecipitated products.

Microanalytical (C,H,N) results indicated the polymers so isolated to be mixtures of the hexafluorophosphate and tetrachloroantimonate salts ($\underset{\sim}{1}$, $A^- = PF_6^-$, $SbCl_4^-$). Persistent hydrolysis (to destroy the $SbCl_4$ complex) and renewed precipitation from aqueous $DMSO-NH_4PF_6$ furnished samples of the pure hexafluorophosphate type. Other samples of the polymers were reprecipitated as the phosphotungstates, but the compositions were not reproducible in these cases.

The IR spectra (KBr) of all polyamides exhibited the typical amide bands (amide I band listed in the table) in addition to the known cobalticenium absorptions and the pertinent benzene-aromatic bands of the respective segments. Carboxyl end group absorption was indicated by a weak band near 1730 cm^{-1}. The ^1H NMR spectra (DMSO), confirming structures $\underset{\sim}{1}$, were characterized by the cyclopentadienyl proton resonances displaying the common A_2B_2 (two peaks, unresolved) and the expected arene proton signals, the arene-to-cyclopentadienyl proton peak intensities being in the correct ratios (Table 1).

TABLE 1. Polycondensation of 1,1'-Dicarboxycobalticenium Chloride with Aromatic Diamines[a]

Ar in Diamine	Time/Temp.[b]	Product Polyamide[c]	Yield,[d] %	Anal. Found,[e] %			^{1}H NMR,δ[f]		Arene/Cp Proton Ratio[g]	Amide I Band,[h] cm^{-1}
				C	H	N	Arene	Cp		
![p-xylylene diamine structure]	12h/155°C, 13h/165°C[i]	1a	66	39.44	3.30	5.39	7.1–7.8[j]	6.4 5.9_5	0.5	1675
![biphenyl diamine structure]	2h/150°C, 13h/170°C, 5h/175°C[k]	1b	49	46.66	3.12	4.35	6.9–8.0	6.6 6.1	1.0	1673
![CH2-bridged diamine structure]	20h/155°C, 8h/160°C	1c	78	50.31	3.80	4.52	6.9–7.8	6.5^l 6.0^l_5	1.0	1673

[a] In SbCl$_3$, equimolar quantities of dicarboxy compound and diamine throughout. Concentration (each reactant), 1.40 ± 0.14 mol l^{-1}. [b] All temps. ± 3°C; 3h heat-up time for temp. interval 100 → 150°C. [c] Color, orange-brown (1a); brown (1b); yellow-brown (1c). [d] Calcd for 1a,b, A$^-$ = PF$_6^-$/SbCl$_4^-$ (50:50); 1c, A$^-$ = PF$_6^-$/SbCl$_4^-$ (90:10). [e] Calcd for 1a: C, 39.51; H, 2.58; N, 5.12; 1b: C, 46.20; H, 2.90; N, 4.49; 1c: C, 50.52; H, 3.39; N, 4.31. Anion composition as given in[d]. [f] In ppm rel. to intern. TMS. Cp = Cyclopentadienyl. [g] Estimated error, ± 10%. [h] On KBr pellets. [i] Similar polymer composition after 32h/155°C, 1h/175°C. [j] Singlet emerging at 7.55 ppm. [k] Similar polymer composition after 22h/165°C, 15h/175°C. [l] Weak signals at 6.8 and 6.2 ppm, suggesting presence of type 2 structures in low concn.

Viscometric investigations on the polyamides proved inconclusive insofar as the measured inherent viscosities, regardless of the solvent systems employed (DMSO, aq. DMSO, N,N-dimethylacetamide, 98% formic acid, 98% sulfuric acid), were in the vicinity of zero and occasionally even assumed negative values indicating a viscosity depression by the solute. Similarly useless results were obtained by vapor pressure osmometry in dimethylformamide. Despite these findings, we assign polymeric structures to the products on the strength of the following two arguments: (i) The materials form films when cast from DMSO solutions onto aluminum foils; the films, while brittle and not self-supporting, adhere well to the substrate when the composite is bent around a 5 mm dia. mandrel. For comparison, similarly prepared composites made up from DMSO solutions of non-polymeric cobalticenium compounds of the previous study (4) lack adhesion and completely fail the bending test, the salt layers crumbling off with ease. (ii) The carboxyl end group absorption at 1730 cm^{-1}, with a peak height generally less than 10% of that of the amide I band, is too weak to account for a monomeric or oligomeric structure 1. Although the presence of non-polymeric cobalticenium structures of the type 2, isomeric with 1 and hence equally consistent with the microanalytical and NMR data, could conceivably cause a low carboxyl end group concentration, such structures must be discarded as major components because of the absence of high-frequency imide carbonyl absorption in the IR spectra and the substantial coincidence of the cyclopentadienyl NMR signals with those in simple cobalticenium diamides (4) with parallel, that is, non-tilted rings.

We are presently investigating the feasiblity of gel permeation chromatography for molar mass determination in an effort to derive a more quantitative measure of the degrees of polymerization attained in these polycondensations. Thermogravimetric analyses will also be performed on the products for further corroboration of their polymeric nature.

III.   EXPERIMENTAL

A.   Solvents and Reagents.

Antimony trichloride, a commercial grade, was purified by

sublimation (60°C/0.01 torr) into the reaction vessel, a
100-ml, two-neck, round-bottom flask equipped as described
(4). Spectrograde DMSO was used for the NMR work. The di-
carboxycobalticenium compound was recrystallized from aqueous
2N HCl. The amines, p-phenylenediamine, benzidine, and meth-
ylene-4,4'-dianiline (4,4'-diaminodiphenylmethane), were recrystallized from deoxygenated aqueous ethanol containing a
pinch of sodium dithionite. All reagents were thoroughly
dried prior to use.

B. Polycondensation Reactions.

The reaction vessel containing the antimony trichloride
(5.0 ± 0.5 g) was charged with 1,1'-dicarboxycobalticenium
chloride (0.624 g, 2.0 mmol) and the amine reactant (2.0 mmol).
After 30 min purging with predried nitrogen at room temperature, the magnetically stirred mixture was heated for one hr
at 100°C under a slow current of nitrogen to achieve a homogeneous solution. Heating was continued with stirring under
N_2 at the temperatures and for the periods tabulated. Small
portions of antimony halide sublimed into the condenser were
occasionally returned to the flask by means of a vertically
sliding glass rod fitted to the condenser top. The cooled
melt was treated in a Soxhlet apparatus over a 24 hr period
with dry ether (3 × 100 ml) for extraction of most of the
antimony trichloride. Subsequently, for polymers 1a and 1c,
the ether-insolubles were exhaustively extracted (SbOCl residues should be grayish-white at this point) with warm (60°C)
water (4 × 80 ml generally sufficient), which dissolved the
product polyamide as a mixture of chloride and tetrachloroantimonate salts. The combined extracts, adjusted to pH 5-7,
were concentrated to 150 ml in a rotary evaporator (30°C) and,
after filtration at 80-90°C and adjustment to pH 4, were
treated with ammonium hexafluorophospahte (0.35 g). The
separated 1, $A^- = PF_6^-$, $SbCl_4^-$, was filtered off after cooling
to room temperature and washed with ice water (10 ml). Concentration of the filtrate to 30 ml, addition of another
0.15 g NH_4PF_6, and standing over night furnished a second
portion of precipitated product salts. The combined portions
were dried for two d at 80°C/0.05 torr. Table 1 lists the
yields and analytical data.

For polymer 1b, the higher-molecular portions of which
were insufficiently soluble in warm water, a somewhat modified
work-up method was employed. To the crude product, after
ether extraction, water was repeatedly added in 5-ml portions
and evaporated in a rotary evaporator (30°C). This resulted
in hydrolysis of admixed $SbCl_3$ and in appreciable (albeit not
quantitative) conversion of $SbCl_4^-$ to Cl^-. The solid evapora-

tion residue was subsequently taken up in DMSO (3 ml), and the filtered solution was stirred into a hot 2.5% aqueous solution of NH_4PF_6 (20 ml). After several hr standing at ambient temperature, the precipitate (1b, $A^- = PF_6^-$, $SbCl_4^-$) was filtered off and washed as before. A smaller second portion was precipitated from the mother-liquor by removal of most of the water in the rotary evaporator and addition of a 3:1 ether/ethanol mixture (30 ml). The combined fractions were dried as described.

All polyamides dissolved readily in DMSO, trifluoroacetic acid, and 98% sulfuric acid, less completely in dimethylformamide, and slightly in ethanol or cold water. They were precipitated from DMSO solution by 3:1 ether/ethanol and incompletely by water.

This work was supported by the South African Atomic Energy Board. The author is indebted to Mrs. Ludmila Bednarik for her invaluable experimental assistance, and to Mr. D. Bylinsky for recording the IR and NMR spectra.

IV.   REFERENCES.

1. Pittman, Jr., C. U., Ayers, O. E., Suryanarayanan, B., McManus, S. P., and Sheats, J. E., Makromol. Chem. 175, 1427 (1974).
2. Szymanski, H. A., Collins, W., and Bluemle, A., Polymer Letters 3, 81 (1965).
3. Saferstein, L., Celanese Research Company, personal communication.
4. Neuse, E. W., and Bednarik, L., Synthesis, in the press.
5. Neuse, E. W., and Horlbeck, G., Polym. Eng. Sci., in the press.

LEAD (IV) POLYESTERS FROM A NEW SOLUTION SYSTEM

Charles E. Carraher, Jr.
Wright State University

and
Charlene Deremo Reese
University of South Dakota

I.   INTRODUCTION

The generation of metal-containing polymers can be a-
chieved using a number of routes. One utilized extensively
by us involves the reaction of an electron deficient site such
as a full cation or the metal atom within a metal-halide
moiety with an electron rich site as typical Lewis
Bases.

The presence of large dipole moments in organolead
halides (1) has been interpreted to indicate a highly ionic
lead-halogen bond. Dialkyllead dihalides are known to react
with a number of Lewis bases, forming the expected condensation products of R_2PbCl_2 with salts of monacids (2,3).

$$R_2PbX_2 \quad + \quad 2RCO_2Na \quad \rightarrow \quad R_2\,Pb(OOR')_2 \quad + \quad 2NaX$$

I.

We have been active in the synthesis of a number of
organometallic polymers via condensation reactions. Much
effort has been aimed at the synthesis of Group IV A containing polymers by condensing Lewis based such as salts of
diacids with diaryl- and dialkyl-dihalo Group IV A compounds
where M = Si, Ge, Sn

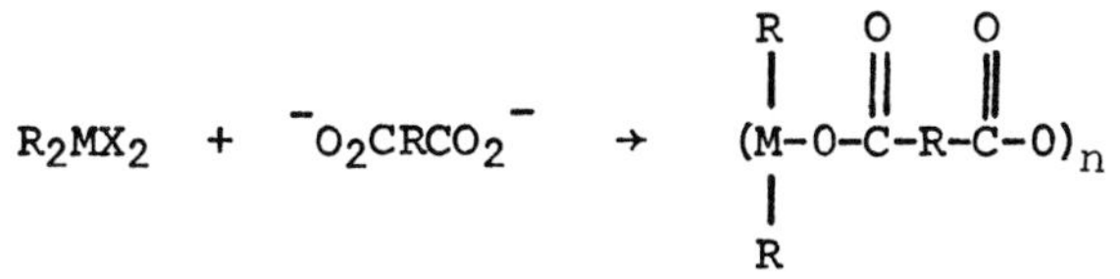

$$R_2MX_2 \quad + \quad {}^-O_2CRCO_2{}^- \quad \rightarrow \quad \begin{matrix} R & O & O \\ | & \| & \| \\ (M\!-\!O\!-\!C\!-\!R\!-\!C\!-\!O)_n \\ | \\ R \end{matrix}$$

II.

but not Pb (for instance 4-6).

A logical extension is the synthesis of analogous lead-
containing products of form I where M = Pb. We now report
this synthesis.

II. EXPERIMENTAL

Perfluoroterephthalic acid (P.C.P., Gainesville, Fl.); tetrafluoroterephthalic acid, tetramethlterephthalic acid (INC labs., Inc., Plainview, N.Y.); adipic acid (J.T. Baker Chemical Co., Philipsburg, N.J.); 2,5-dichloroterephthalic acid, bromoterephthalic acid, nitroterephthalic acid, terephthalic acid (Aldrich Chemical Co., Inc., Milwaukee, Wis.); diethyllead dichloride, dibutyllead dichloride and diphenyllead dichloride (Alfa Inorganics, Inc., Beverly, Mass.) were utilized as received.

Reactions were carried out in a one quart Kimex emulsifying jar placed on a Waring Blendor (Model 1120). Blendor speed was controlled using a Type 116 Powerstat. The lead-containing solution was added through a powder funnel inserted into a hole in the cover of the jar to the stirred acid containing solution.

After stirring was stopped, the contents of the emulsifying jar were filtered under suction and the product obtained as a white, cakey material. Suitable washing and drying procedures were then employed.

Infrared spectra were obtained using KBr pellets with the Perkin-Elmer 170A and 237B and Beckman IR-10 Spectrophotometers. Elemental analysis was carried out using both a calcium oxide fusion and du Pont 950 Thermal Gravimetric Analyser. Both spectral studies and elemental analysis are in agreement with the structure given in I where M = Pb.

III. DISCUSSION

We have synthesized a number of metal containing polymers via condensation reactions, generally using interfacial or solution techniques. For many desired syntheses the interfacial technique is the method of choice for several reasons, particularly because it is often a more favorable thermodynamic system and usually is capable of generation of products from less reactive co-reactants compared with solution techniques. Interfacial systems do require that an interface be formed. This is normally accomplished by using an aqueous phase and an organic phase largely immiscible with the acqueous phase.

Previous attempts by us at the synthesis of lead-containing polymers have failed because of the insolubility of suitable lead-containing reactants in systems suitable either as aqueous phases or as organic phases. For instance, diethllead dichloride was insufficiently soluble in common organic solvents such as benzene, p-chorotoluene, 2,5-hexanedione, xylene, benzonitrile, heptane, acetonitrile, methyl ethyl ketone, and ethyl acetate to prepare 0.25 mmole per 25 ml solutions (usually found to be within the lower limits for successful interfacial and organic solution systems).

The lead compounds were soluble in 2,5-hexanedione and acetonitrile solutions saturated with LiBr or LiCl (0.5 mmole /25 ml). No polymer was produced when these were mixed with aqueous solutions containing salts of the diacids. After a number of other unsuccessful attempts at polymer formation, it was found that organometallic compounds were sufficiently soluble in DMSO and DMF to give 1.00 mmole/25 ml solutions. After further work, our attempts became aimed at the use of DMSO containing systems. When water was poured directly into a solution of 1.00 mmole of Et_2PbCl_2/25 ml DMSO, the lead monomer precipitated out; but, addition of a solution of 50% DMSO-H_2O [v/v] did not produce precipitant. This was important since   a. presence of water is necessary to produce the active carboxylate anion;   b. polymer is normally recovered as the precipitant in reaction mixtures; c. the diacids are insoluble in DMSO but the diacid and its disodium salt are soluble in a 50-50 mix of DMSO-H_2O. This system did permit synthesis of the desired lead (IV) poly-esters. (Table I).

Table 1. Results as a function of reactants.

Aromatic Diacids	Yield[a] (ET PbCl) (%)	Yield[b] (Ph PbCl) (%)	Hammett[c] σ
Perfluoroterephthalic Acid	4	trace	1.36
2,5-Dichloroterephthalic Acid	30	12	0.74
Nitroterephthalic Acid	32	--	0.71
Bromoterephthalic Acid	44	--	0.39
Terephthalic Acid	67	24	0.00
2,5-Dimethylterephthalic Acid	75	27	-0.14
Tetramethylterephthalic Acid	62	--	=0.28
Adipic Acid	86	--	
Sebacic Acid	--	20	
Malonic Acid	--	5	

[a]Reaction conditions: diethyllead dichloride (1.00 mmole) in DMSO (25 ml) added to stirred (21,000 rpm no load) solutions of terephthalic acid (1.00 mmole in 50% DMSO-H_2O [v/v]) with sodium hydroxide (2.00 mmole) at 23°C **and noted stirring times.**

The products are insoluble in all solvents tested including the following solvents: hexamethylphosphoramide,
dimethyl sulfoxide, Sulfolane-W, 2,5-hexanedione, acetonitrile, triethyl phosphate, and diethyl malonate. Tests were
conducted at both 25 and 80°C over a two week observation
period.

This insolubility is consistant with the previously
observed solubility trends where solubility is of the order
Si Ge Sn for any given Group IV A containing polymer where
tin products are generally soluble in only a limited number
of dipolar aprotic solvents or insoluble in all attempted
solvents (6,7).

Yield and product molecular weight are the two most
utilized parameters for the evaluation of reaction variables.
Because of the insolubility of the products (preventing
molecular weight determination) only yield will be used as
the exhibitive parameter.

Reaction is fast, presumably complete in about 20 seconds. The product appears stable in the reaction mixture for
about 2 minutes. This system is highly reactive since it is
capable of giving a high yield rapidly. Representative
results appear in Table 2. Results given in Table 1 then are
"maximum" results rather than "kinetic" results.

Table 2.   Results as a function of stirring time.[a]

Time (sec)	Yield (%)
10	58
15	61
20	65
30	67
45	68
60	67
120	66

[a] Same as Table 1a, where the acid is terephthalic acid.

b.   Ibid a. except using diphenyl lead dichloride.

c.   I. March, Advanced Organic Chemistry; McGraw Hill, N.Y.,
     1966, p. 241.

Reaction does appear general, occurring with both aliphatic and aromatic diacids and dialkyl and diaryl lead di-
chlorides.

Aliphatic silanes and stannanes react to give a greater
yield with a given carboxylate anion than do the analogous
aromatic compounds (4). The same is found for plumbanes
(Table I).

π-Bonding [(d-p)] is felt to occur between the phenyl-
π-electrons and empty "d"-orbitals on the metal. Such (d-p)π
bonding as in Form III is invoked to explain many kinetic
and spectral trends involving Group IV A organometallic
compounds (8).

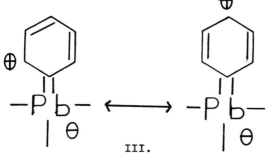

III.

Such (d-p)π bonding may cause a reduction in the
electronegativity of the metal atom and/or a reduction of the
availability of the "d"-orbitals to participate in accepting
the carboxylate ion via an associative reaction pathway.
Either or both of the above would result in a decrease in
polymerization rate and may be responsible for the observed
yield trend.

Table 1 also presents yield as a function of substituted
terephthalic acid together with their associated Hammett σ
coefficient. Thus the more nucleophilic acid salt is favored
in the reaction with the electron poor lead chloride moiety.
This is consistant with a single nucleophillic attack on the
lead atom as being the critical (rate determining) step.
Condensation with tetramethylterephthalic acid does not
follow this general trend probably due to the greatly
increased steric factors present.

IV.   REFERENCES

1. G. Lewis, P. Oesjer, and C. Smyth, J. Am. Chem.
    Soc., 62, 3243 (1940)
2. F. Huber, M. Enders, and R. Kaiser, Z. Naturforsch,
    21b, 83 (1966).

3. F. Huber and M. Enders, Z. Naturforsch, 20b, 601 (1965).

4. C. Carraher and R. Dammeier, Makromolekulare Chemie, 135, 197 (1970); 141, 245 (1971); 141, 251 (1971); 141, 259 (1971).

5. C. Carraher and R. Dammeier, J. Polymer Sci., A-1, 8, 3367 (1970); 10, 413 (1972).

6. C. Carraher, Inorganic Macromolecules Reviews, 1, 271 (1972).

7. C. Carraher and F. Millich, Eds., Interfacial Synthesis, Dekker, N.Y. (1976).

8. T. Birchall and W. Dolly, Inorganic Chem., 5, 2177 (1966).

The authors are pleased to acknowledge support from American Chemical Society - Petroleum Research Foundation Grants 9126-Bl, 3-C and 7814 - Bl,3.

REACTION OF TRIPHENYLANTIMONY DICHLORIDE WITH POLYACRYLIC ACID

Charles E. Carraher, Jr.
Wright State University

and

Mark Moran
University of South Dakota

I. INTRODUCTION

We have been active in the modification of polymers by (for instance 1-11), including metal-containing moieties into the side chains of the polymers. We have based much of the emphasis in this area on a number of themes, including the following: A. The use of commercially available polymers better signals the potential industrial application of our techniques, because of the ready availability of the reactants and low cost of at least the polymeric component. B. Significant changes of the properties of the products generally results from these modifications. For instance, polyvinyl alcohol degrades > 80% before 300°C whereas inclusion of ca 10 mole-% triphenyl tin groups permits a product where only ca 20% weight loss occurs to 800°C in both air and nitrogen. Many of the utilized polymers are water soluble whereas after modification they are hydrophobic. Most of the polymers exhibit bulk resistivities greater than 10^{10} ohm-cm whereas modified materials exhibit resistivites often in the range of 10^4 - 10^9 ohm-cm. C. Such modifications are intended to both test and to illustrate the relative reactivity of functional groups contained within a polymer versus those in smaller molecules.

Most of our modifications can be rationalized utilizing Lewis acid-base concepts. We have thus far effectively modified materials such as polyvinyl alcohol (1,2,4,8), polyacrylic acid (6,9), polyethleneimine (11) and poly-acrylonitrile (3,5,7,10). While we have completed many modifications, we have not studied such modifications as a function of polymer molecular weights. We now report the initial modification of polyacrylic acid with antimony (V) halides as a function of a number of reaction variables including the molecular weight of polyacrylic acid.

II. EXPERIMENTAL

I.   II.

Triphenylantimony dichloride (Alfa Inorganics, Beverly, Mass.) and the polyacrylic acids (Polysciences, Inc., Warrington, Pa.) were used as received.

Reaction procedures are similar to those described in detail elsewhere (16). Briefly the antimony-containing organic phase is added to stirred aqueous phases containing polyacrylic acid (PAA) and added base. Modified PAA rapidly precipitates from the reaction mixture and is separated by suction filtration. Since some unreacted PAA may be included within the precipitate, the recovered solid is added to a beaker containing ca 250 ml water and set aside for two days, whereupon it is again filtered. The filtrate contains unreacted PAA as determined by IR spectroscopy. The wash filtrate is added to the original aqueous filtrate. (The reaction mixture is made acidic by addition of dilute HCL after the product is removed by filtration.) The aqueous portions are evaporated to dryness, washed quickly and filtered. The insoluble material is unreacted PAA (by infrared analysis) which takes several hours to go into aqueous solution. Amount of antimony moiety inclusion is calculated from knowledge of the unreacted PAA. This agrees within 3% with that determined by typical wet analysis.

IR spectra were obtained utilizing KBr pellets and a Perkin Elmer 237B spectrophotometer. Spectra were consistent with a product of form I or/and II.

III. RESULTS AND DISCUSSION

There are a number of reasons for desiring the inclusion of antimony into PAA chains. These include potential medical-biochemical applications. We previously synthesized a number of arsenic containing polypyrimidines which appeared to

be specific toxic control-release agents. Organic antimonials are believed to behave similarly (at least in select
situations) to the arsenicals (12).

Antimony potassium or sodium tartrate, stibophen, sodium
antimonyl gluconate and sodium-α,α'-dimercaptosuccinate are
used to control filariasis, leishmaniasis and schistosomiasis
(12,13). Their high toxicity is a disadvantage to human
applications. Formation of a mixed chelate of antimony
sodium tartrate with penicillamine gives a nontoxic material
that retains its antiparasitic action in schistosomiasis (14).
Connection of organic antimonials through a polar, potentially
biologically degradable moiety (here the carboxyl grouping)
might result in the formation of medically useful control
reagents.

Addition of antimony to the PAA chain should also bring
about a number of potentially useful physical changes. These
should include a marked increase with respect to thermal
stability, different solubility properties, etc.

Initial and subsequent attempts at the inclusion of
monohaloantimony (V) compounds to produce linear products of
form III were not successful, but attempts at the inclusion
of the analogous dihalo reactants were successful. Thus,
additional efforts were aimed at the inclusion of R_3AsX_2
reactants. Here we will consider only the reaction of triphenylantimony (V) dichloride with PAA.

$$
\begin{array}{ccccc}
\text{H} \;\; \text{H} & & & & \text{H} \;\; \text{H} \\
-\text{C}-\text{C}- & + & R_4SbX & \rightarrow & -\text{C}-\text{C}- \\
\text{H} \;\; \text{C}=\text{O} & & & & \text{H} \;\; \text{C}=\text{O} \\
\;\;\; \text{O}^{\ominus} & & & & \;\;\; \text{O} \\
& & & & \;\;\; \text{SbR}_4
\end{array}
$$

III.

Percentage product yield depends on the structure of the
resulting product which is probably a combination of forms I
and II, the proportion of which will vary from product to
product. There appears to be an absence of Sb-OH groups
(which can be partially hid by the C-OH moiety). While wet
and IR spectral analysis indicate an absence of Sb-Cl groups,
neither is sufficiently sensitive or well established to pre-
clude the presence of such groups. Results were generated
assuming a product of form I.

Reaction is rapid with both yield and antimony inclusion
reaching a maximum after 60 secs stirring time. Decrease in
yield and antimony inclusion at longer stirring times is

typical for other antimony (like antimony polyesters, polyamines) systems studied by us and is due to base catalyzed hydrolysis of modified product (for instance 15-17).

Table 1.   Results as a function of stirring time and PAA molecular weight.

	Molecular Weight ($\bar{M}_n$)					
	8.0×10^4		1.5×10^5		2.5×10^5	
Stirring Time (secs)	Yield (%)	Antimony Inclusion (%)	Yield (%)	Antimony Inclusion (%)	Yield (%)	Antimony Inclusion (%)
15			17	1		
30	14	16	20	12	20	9
60	23	44	42	45	74	74
120	16	23	18	28	29	29
300			16	11	20	18

Reaction conditions: Chloroform solutions (25 ml) containing triphenylantimony dichloride (1.00 mmole) are added to rapidly stirred (21,400 rpm no load stirring rate) aqueous solutions (25 ml) containing PAA (1.00 mmole) and sodium hydroxide (1.00 mmole) at 25°C.

Comparison of results as a function of PAA molecular weight is more difficult to assess. Generally reaction is more complete with respect to both yield and antimony inclusion as PAA molecular weight increases. Explanations include invoking a type of auto (or related) catalytic effect, and greater occlusion of antimony co-mer prior to product precipitation by higher molecular weight chains. If the latter were the dominant factor, one might expect both yield and/or antimony inclusion to increase as the solubility of PAA (or its salt) in the organic phase increases. Neither yield nor antimony inclusion exhibits a trend with respect to PAA solubility (Table 2, solubility of the salt of PAA increases from top to bottom being low [ca less than 0.01 g/100 ml] throughout). Thus, PAA solubility alone is not the critical factor. As a further note, reaction probably occurs near the interface or within the aqueous phase thus minimizing the effects of PAA solubility in the organic phase.

Currently, it is best to simply cite trends with respect to PAA molecular weight. With respect to all reaction variables studied yield and antimony inclusion remained essentially the same or increased as PAA molecular weight

increased, with general trends independent of PAA molecular
weight. Thus only results for the middle molecular weight
PAA will be subsequently reported.

As seen in Table 2, the nature of the organic solvent is
important, probably exhibiting its control by regulating antimony flow to the reaction site and/or similar interfacial
tension and/or distribution ratio related phenomena.

Table 2. Results as a function of organic solvent

Organic Solvent	Yield (%)	Antimony Moiety Inclusion (%)
Carbon Tetrachloride	72	87
Benzene	98	98
Toluene	48	42
Chlorobenzene	53	38
Chloroform	42	46
Nitromethane	56	32

Reaction conditions. Same as Table 1 except utilizing one
min stirring times, PAA of M_n = 1.5 X 10^5 and the described
organic solvent.

The use of a number of bases is possible. If base only
acts to neutralize the PAA then reaction should be independent of the nature of the base. Base is necessary for modification to occur (Table 3) in agreement with the need for formation of the carboxylate anion, but yield is dependent on the
nature of the base, indicating that added base plays a role(s)
in addition to that of neutralizing the PAA. Furthermore,
yield increases as the amount of base added increases, as illustrated in Table 4 for sodium hydroxide. Reasons for the
afore dependencies are currently of a speculative nature.

Table 3. Yield as a function of base type

Base	Yield (%)	Antimony Moiety Inclusion (%)
NaOH	42	45
NH_3	11	20
Et_3N	6	11
None	0	–

Reaction condition: Ibid Table 2 except for base used using
chloroform as the organic solvent

Table 4. Yield as a function of the amount of base.

Amount of Base (moles $\cdot$ 10^3)	Yield (%)	Antimony Moiety Inclusion (%)
0.50	15	23
1.00	42	45
2.00	51	70

Reaction conditions: Ibid Table 2 except for amount of socium hydroxide used with chloroform as the organic solvent.

Reaction is "diffusion controlled" throughout the stirring range studies (Table 5). This is consistent with the previously high reactivity of carboxylate anions to Group V organometallic halides (for instance 16) observed.

Table 5. Yield as a function of stirring rate.

Stirring Rate (rpm)	Yield (%)	Antimony Moiety Inclusion (%)
6,800	12	4
12,700	14	5
17,100	19	14
21,400	42	45
23,400	45	48

Reaction conditions: Ibid. Table 2 except for stirring rate using chloroform as the organic solvent.

Yield maximizes at a 0.02 molar concentration of reactants (Table 6), but exhibits a complex behavior with respect to monomer ratio (Table 7). There is an opposing trend with respect to yield and antimony inclusion. Yield increases but antimony inclusion decreases as the amount of PAA increases; whereas yield decreases but animony inclusion increases as the amount of ϕ_3SbCl_2 increases. Neither trend is unexpected, both pointing out the dependence of yield on PAA and indpendence of antimony inclusion on ϕ_3SbCl_2. The precise interrelationship between the reactants and reactant ratios though is not known.

Table 6.   Yield as a function of reactant concentration.

Reaction Amount (moles · 10^3)	Yield (%)	Antimony Moiety Inclusion (%)
0.25	5	2
0.50	19	41
1.0	42	45
2.0	31	14

Reaction conditions: Ibid. Table 2 except for amount of reactants using chloroform as the organic solvent.

Table 7.   Yield as a function of monomer ratio of reactants.

Monomer Ratio (ϕ_3SbCl_2:NaOH:PAA)	Yield (%)	Antimony Moiety
1:0.5:0.5	36	80
1:1:1	42	45
1:2:2	78	27
0.5:1:1	61	36
1:1:1	42	45
2:1:1	22	72

Reaction conditions: Ibid. Table 2 except for molar ratios of reactants where chloroform is the organic solvent.

As an additional note, high substitution is not needed to achieve high yield (Table 6). This may speak against any previously mentioned "auto catalytic" or related effect.

IV.   REFERENCES

1.   C. Carraher and J. Piersma, J. Macromol. Sci.-Chem., A7, 913 (1973).

2.   C. Carraher and J. Piersma, Angew, Makromolekulare Chemie, 28, 153 (1973).

3.   C. Carraher and L-S Wang, Angew. Makromolekulare Chemie, 25, 121 (1972).

4.   C. Carraher and L. Torre, Angew. Makromolekulare Chemie, 21, 207 (1972).

5.   C. Carraher and L-S. Wang, J. Polymer Sci., A-1, 9, 2893 (1971).

6.   C. Carraher and J. Piersma, J. Applied Polymer Sci., 16, 1851 (1972).

7. C. Carraher and L-S. Wang, J. Macromol. Sci-Chem., A7, 513 (1973).

8. C. Carraher and L. Torre, J. Polymer Sci., A-1, 9, 975 (1971).

9. C. Carraher and J. Piersma, Makromolekulare Chemie, 152, 49 (1972).

10. C. Carraher and L-S. Wang, Makromolekulare Chemie, 152, 43 (1972).

11. C. Carraher and M. Feddersen, Angew. Makromolekulare Chemie, 54, 119 (1976).

12. E. Bueding and J. Fisher, Biochem. Pharmacol., 15, 1197 (1966).

13. J. Casals, Brit. J. Pharmac., 46, 281 (1972).

14. M. Pedrique and N. Freoli, Bull. World Health Org., 45, 411 (1971).

15. C. Carraher, W. Moon and T. Langworthy, Polymer Preprints 17, 1 (1976).

16. C. Carraher, J. Sheats and H. Blaxall, Polymer Preprints, 16, 655 (1975).

17. C. Carraher and W. Moon, Eup. Polymer J., in press.

We gratefully acknowledge support of this research by the American Chemical Society Petroleum Research Foundation grant numbers 7814-Bl, 3 and 9126-Bl, 3-C.

PHOTOCROSSLINKABLE ORGANOMETALLIC POLYESTERS

Douglas G. Borden

Research Laboratories, Eastman Kodak Company

ABSTRACT

Metals may be incorporated into polyesters by solution polymerization of a mixture of bisphenols and diacid chlorides with a portion of the acid chloride replaced by an organometallic dihalide. If one of the bisphenols is DVCP [2,5-bis(4-hydroxy-3-methoxybenzylidene)-cyclopentanone], the polymer will be crosslinkable on exposure to light. DVCP and either tetrachlorobisphenol A or tetrabromobisphenol A were polymerized with sebacyl chloride and one of 27 organometallic dihalides containing one of 17 metals (antimony, arsenic, boron, germanium, hafnium, iron, lead, manganese, palladium, platinum, ruthenium, selenium, silicon, tin, titanium, vanadium, and zirconium). The polymerization was run in a mixture of 1,2-dichloroethane and 1,1,2-trichloroethane with triethylamine as the catalyst and acid acceptor. The high molecular weight, soluble polyesters obtained were coated on aluminum or copper, exposed imagewise to ultraviolet and visible light, and solvent developed to produce a photoresist incorporating the metal in the polymer. From 0.5% to 20% of the polymeric image was the metal.

I. INTRODUCTION

Since the discovery of ferrocene, organometallic chemistry has expanded in many directions. Now available are organic compounds that contain metals in their structures and yet are soluble in organic solvents. Many early organometallic polymers were insoluble, and others which retained some organic solvent solubility were very low in molecular weight and so did not have the properties of high polymers. Carraher et al. (1-12) prepared many organometallic condensation polymers by interfacial polymerization. Others prepared polymers from metallocenes (13, 14). Many of these polymers are low molecular weight or insoluble. The challenge of preparing and studying high molecular weight polymers with

metals in the polymer backbone led us to prepare a series of
light-sensitive polyesters containing organometallic units.
Light-sensitive polyesters which are crosslinkable on exposure
to light are used in photolithography and as photoresists.

We have been particularly interested in condensation
polymers prepared from DVCP. This bisphenol, when incorporated into a polymer, has provided materials which could be
photoinsolubilized with exposure to light less than that
required by the sensitized poly(vinyl cinnamates).

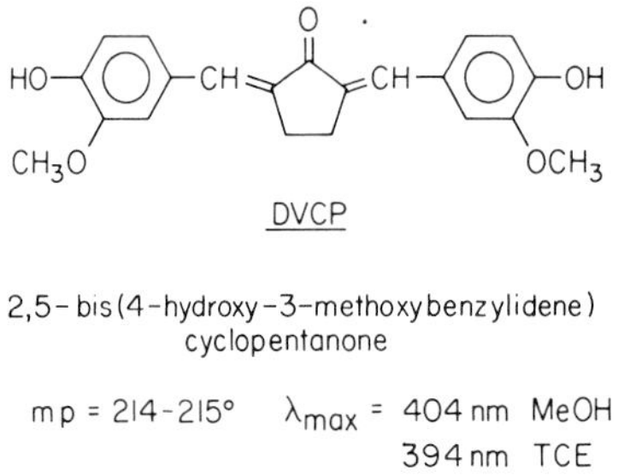

DVCP

2,5-bis(4-hydroxy-3-methoxybenzylidene)
cyclopentanone

$mp = 214-215°$ $\lambda_{max} = 404\,nm$ MeOH
                             $394\,nm$ TCE

Fig. 1.

Polycarbonates of DVCP and phosgene are soluble only in
tetrachlorethane, but those of DVCP and a bischloroformate
such as those from neopentyl glycol or nonanediol are soluble
in 1,2-dichloroethane (DCE) or in 1,1,2-trichloroethane
(TCE). The solubility of these polycarbonates was greatly
improved and the light sensitivity was not greatly diminished
by replacing up to 50 mole% of the DVCP with a non-light-
sensitive bisphenol such as bisphenol A, tetrachlorobisphenol
A, or tetrabromobisphenol A (Figure 2).

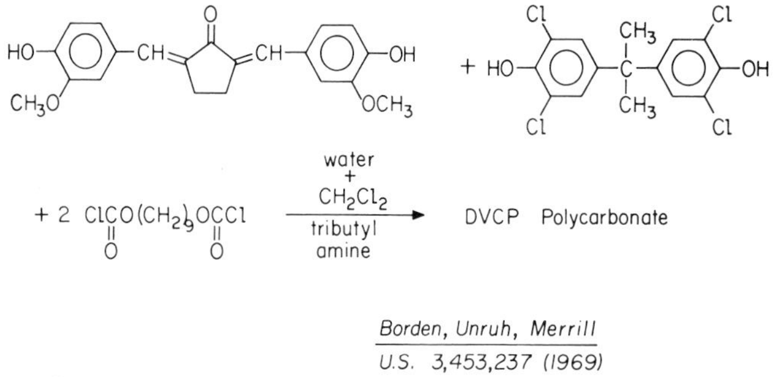

$$+ 2\ ClCO(CH_2)_9OCCl \xrightarrow[\substack{tributyl \\ amine}]{\substack{water \\ + \\ CH_2Cl_2}} DVCP\ \ Polycarbonate$$

Borden, Unruh, Merrill
U.S. 3,453,237 (1969)

Fig. 2.

DVCP polysulfonates prepared both in solution and in
interfacial condensation by Arcesi et al. (16, 17) were
remarkably resistant to strong caustic etchants. The corresponding phosphorus-containing polymers of DVCP were prepared
in solution and were good photoresists in acid etchants (18).

This solution polymerization technique was used to
prepare organic-soluble DVCP polyesters with weight average

molecular weights >100,000 (19).   Although many acid chlorides
were used, the best polyesters were obtained from azelaoyl
chloride, sebacyl chloride, or isophthaloyl chloride.

Replacing up to 50 mole% of the acid chloride with
various organometallic dihalides produced organometallic
polyesters which were still photocrosslinkable.   This provided
a means of forming photographic images with a uniform amount
of metal in the polymeric image.   Two series of organometallic
DVCP polysebacates were prepared to determine the maximum
amounts of organometallic intermediates which can be incorporated into the polyesters while still retaining the organic
solubility and photosensitivity of the nonorganometallic
polymers.

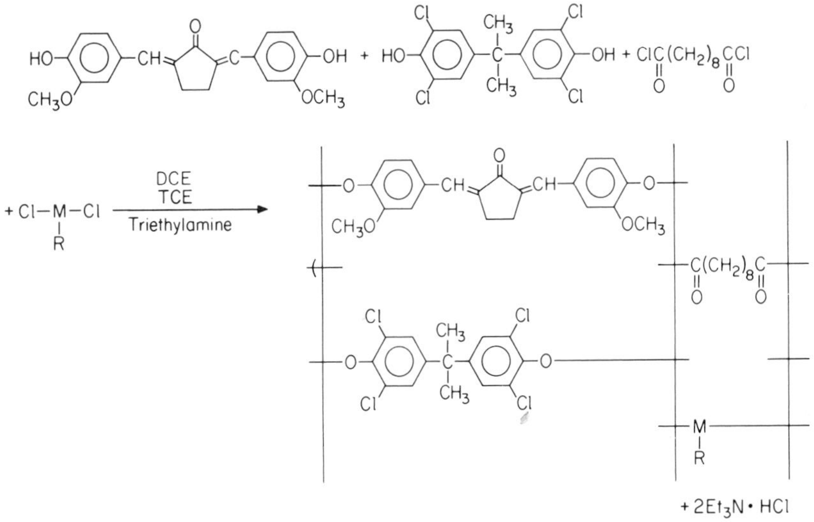

Fig. 3.   Organometallic DVCP Polysebacate.

The 27 commercially obtained organometallic dihalides
contained one of 17 metals.   In the first series of polymers
(Table I), a mixture of 56 mole% of DVCP and 44 mole% of
tetrachlorobisphenol A was reacted with sebacyl chloride
having a certain amount replaced by the organometallic
dihalide.   Table I shows the mole% of organometallic dihalide
in the monomer feed and also the theoretical and actual
amounts of metal in the polymer.   The metal analyses by
flame spectroscopy were made difficult by the flame-extinguish-
ing properties of the preferred chlorinated solvents.

Table II shows the second series of polyesters, in which
a mixture of either 65:35 or 60:40 mole% of DVCP and tetrabromobisphenol A was used with sebacyl chloride and the
organometallic dihalide.

The amount of organometallic compound which could be incorporated into a soluble polyester ranged from about 0.5% to 50% of the acid moiety. Polysebacate-siloxanes were prepared with 12-50% of the dichlorosilane in the reaction mixture but most of them had viscosities of less than 0.22 indicating that a chain-terminating process such as hydrolysis of the dichlorosilane or of the chlorosilane oligomer occurred. The actual silicon content of the polymer was close to theoretical. With the tin polyesters, 12-25 mole% of the tin dihalide was used in the feed but analysis showed less than half of the tin expected. Tables I and II show the amount of organometallic intermediate used and the amount found in the polymer.

The polysebacates of DVCP usually precipitate as long, tough yellow fibers, but polymers having a viscosity less than 0.40 precipitate as granules or powder.

DVCP polyesters have previously been prepared and it was easy to obtain a high molecular weight polymer with a corresponding high degree of photosensitivity. A footnote on Table I explains the determination of a qualitative "speed" or photosensitivity value. The polyesters with the organometallic groups present did not show any significant shift in UV or visible absorption except that from ferrocene dicarbonyl chloride which had some absorption in the 550 nm region of the spectrum. There was no appreciable sensitization or desensitization of the DVCP polyester noted which might be due to the organometallic unit.

Example 1 is a typical preparation of an organometallic DVCP polyester. Example 2 describes the preparation and evaluation of a photoresist from a DVCP-hafnicinium polyester.

II.   EXPERIMENTAL

A.   Reactants

The commercially obtained organometallic dihalides were used as received from the manufacturers except triphenylarsenic dichloride which was recrystallized from DMF-ether providing white crystals melting at 168°.

Phenylboron dichloride, di-n-butylgermanium dichloride, dimethylgermanium dichloride, diphenylgermanium dichloride, diphenyl lead dichloride, bis-(triphenylphosphinatopalladium) dichloride, tris-(triphenylphosphinatoruthenium)dichloride, diphenylselenium dichloride, dibenzyltin dichloride, di-n-butyltin dichloride, dicyclohexyltin dibromide, dimethyltin dichloride, dioctyltin dichloride, diphenyltin dichloride, diacetylacetonyltitanium dichloride and zirconocene dichloride were obtained from Alfa Products; triphenylantimony dichloride, dimethyldichlorosilane, diphenyldichlorosilane, and titanocene dichloride from Eastman Organic Chemicals; triphenylarsenic dichloride from Columbia Organic Chemicals; hafnicinium dichloride from Strem Chemicals; ferrocene-1,1'-dicarbonyl-chloride, dipyridylmanganese dichloride, bis-(triphenyl-phosphinatoplatinum) dichloride from Research Organic/Inorganic; methylphenyldichlorosilane from M&T; and zirconocene dichloride from Aldrich Chemicals.

Tetrachlorobisphenol A mp 132-4° was obtained from Dover Chemical Co., tetrabromobisphenol A mp 182-4° from Dow Chemical Co., and sebacyl chloride, bp 181-3° (16 mm) from Eastman Organic Chemicals.

DVCP was prepared by the condensation of vanillin and cyclopentanone in absolute alcohol with borontrifluoride etherate as the catalyst. Recrystallization from 1,1,2-tri-chloroethane gave yellow crystals, mp 214-215°; UV absorption peaks at 404 nm (MeOH), 393 nm (TCE); infrared maxima at 856, 908, 936, 1000, 1035, 1130, 1175, 1215, 1270, 1450, 1520, 1600, and 1675 cm^{-1}; MS = 352, 351 (-H), 337 (-CH$_3$), 335 (-OH), 321 (-OCH$_3$), 228, 176, 160, 137, 119, 91; NMR (in DMSO) relative to TMS, δ2.52 ppm = methylene protons on cyclopenta-none ring, 3.35 ppm = methoxy methyls, 6.72 ppm = olefinic protons, 6.65 ppm = aromatic proton multiplet.

1. Example 1. Solution Polymerization of DVCP, Tetrachloro-bisphenol A, Sebacyl Chloride, and Diphenyl Lead Dichloride (56:44-88:12)

A mixture of 9.87 g (0.028 mole) of DVCP and 8.05 g (0.022 mole) of tetrachlorobisphenol A in 150 ml of DCE and 100 ml of TCE (both dried over molecular sieves) was placed in a 500-ml 3-necked flask fitted with a paddle stirrer, a reflux condenser, and a dropping funnel. The mixture was stirred at room temperature at about 250 rpm and 12.25 g (0.1125 mole) of freshly distilled triethylamine was added. The reaction mixture immediately turned bright red with the formation of the phenolate ion of DVCP. Over a 2-min period, a slurry of 10.52 g (0.044 mole) of sebacyl chloride and 2.6 g (0.006 mole) of diphenyl lead dichloride in 50 ml of mixed DCE and TCE was added with stirring. The temperature rose to about 35°C, and the lead compound dissolved as it reacted. On completion of the addition, the reaction turned from red back to yellow, and the triethylamine hydrochloride began to precipitate. The reaction was heated further at reflux (92°) for 2 hr with stirring. After chilling the clear solution in ice water to 5°, a trace of acetic acid was added and the amine salt was filtered off. The polyester was precipitated by pouring the dope, as a fine stream, into 4 liters of ligroine (bp 30-60°). The long yellow fibers were then slurried in fresh ligroine, then in water and in methanol in a Waring Blendor to remove the last of the salt.

After filtering and washing with water and again with methanol, the polymer was dried at 50° in a circulating-air oven. The yield was 29 g of yellow fibers. $\{\eta\}$ = 0.60 in 1:1 phenol: chlorobenzene. Anal. Calc'd: Cl, 11.4, Pb, 4.5. Found: Cl, 9.8, Pb, 3.4. The polymer was soluble in DCE and had a "speed" or sensitivity value of 10,000. A wedge spectrogram prepared from this polymer showed crosslinking and insolubilization with exposures from 270-460 nm. The peak UV absorption was at 366 nm (run in DCE using a Cary model 15 spectrophotometer).

2. Example 2. Resist Evaluation of a Polyester Prepared from DVCP, Tetrabromobisphenol A, Sebacyl Chloride, and Hafnocinium Dichloride (60:40-88:12) $\{\eta\}$ *= 0.27 dl/g*

A 5-wt% solution of the hafnocene-containing polymer in TCE was whirl-coated on a clean 1.5-mil copper-clad circuit board and the coating was dried for 20 min at 50°. The board was then exposed through a line negative for 2 min in a Colight XPOSER printer which contains a high-pressure mercury source. The board was tray developed in a mixture of DCE and TCE. It was then post-baked for 30 min at 50°. It was spray etched in a 42° Baume ferric chloride solution at 130°F for 6 min. All but the finest lines and dots of the test pattern held up through the etching. The resultant circuit board was comparable to one prepared from a DVCP polyester with no organometallic element present. This shows the feasibility of producing polymer images with a uniform distribution of a metal in the polymer.

3. Example 3. GPC Fractionation of Polymer from DVCP, Tetrabromobisphenol A, Sebacyl Chloride, and Diphenyl Lead Dichloride (65:35-88:12) $\{\eta\}$ *= 0.60 dl/g*

The polyester was fractionated in methylene chloride by preparative GPC and the fractions were analyzed by atomic absorption. A total charge of 1 g of polymer provided only 0.748 g in the eluent. The lead was highly concentrated in the lowest molecular weight fractions, suggesting that a lead chloride group on an oligomer might hydrolyze and give early chain termination.

Fraction	Polymer (g)	Wt% Pb
1	0.176	0.07
2	0.247	0.038
3	0.181	0.03
4	0.049	1.40
5	0.095	1.77
Total	0.748	

4.   Example 4.   GPC Fractionation of Polymer from DVCP, Tetra-bromobisphenol A, Sebacyl Chloride, and Diphenyl Tin Dichloride (65:35-88:120) $\{\eta\} = 0.52 \; dl/g$

This tin polymer was fractionated as in Example 3. Atomic absorption analyses are:

Fraction	Polymer (g)	Wt% Tin
1	0.097	0.4
2	0.293	0.2
3	0.164	0.3
4	0.172	0.3
5	0.007	6.0
6	0.186	0.2
7	0.0036	11.0
Total	0.923	

As in Example 3, the metal was concentrated in the lowest molecular weight fractions.

III.   RESULTS AND DISCUSSION

The solution polymerization of two bisphenols, DVCP and a halogenated bisphenol A, with sebacyl chloride gave a surprisingly versatile means of incorporating organometallic groups into a rapidly growing polymer. Replacing up to 50% of the sebacyl chloride with an organometallic dihalide led to polyesters containing the metal in the polymer backbone while retaining the organic solvent solubility and photo-sensitivity of the corresponding nonmetallic polymer.

The maximum amount of organometallic intermediate which could be used in the monomer feed and still provide high molecular weight polymers has not been determined, but Tables I and II, which summarize all the polymers, show that usually 12-15% or even 20-25% has little effect on the polymer build-up. Hydrolysis of the organometallic dihalide is probably the greatest problem. When phenylboron dichloride was used, it was necessary to add the boron compound from a sealed vial with a syringe directly to the polymer reaction mixture.

The organometallic moiety in the amounts present in these polymers did not sensitize or desensitize the light-sensitive polymers. Exposure of a lead-containing polyester to electron-beam radiation caused crosslinking in the exposed areas which would be characteristic of polymers with the extensive conjugation present. It is possible that organo-metallic polyesters of the type described might be useful in imaging systems using X-rays or other energy sources where the high cross-section of the metals such as lead would be significant.

TABLE I

Organometallic Polysebacates Prepared From DVCP and Tetrachlorobisphenol A

Metal	Organic Ligand	Mole % Metal Interm. in Feed	Polymer Analyses					
			Wt % M Calc.	Wt % M FD.	Cl Calc.	Cl FD.	$\{\eta\}$	Speed*
Tin	Dimethyl	15	3.4	2.4	10.9	12.3	0.24	65
	Di-n-butyl	20	4.4	0.2	10.6	11.6	0.28	90
	Dicyclohexyl	20	4.3		11.4		0.22	40
	Diphenyl	15	3.3	3.4	10.5	10.5		100
	Dibenzyl	20	4.3	2.2	10.4	10.8	0.15	50
	Dioctyl	20	4.2	1.1	10.1	10.1	0.19	70
Lead	Diphenyl	12	4.5	0.3	11.4	11.8	0.66	6500
Boron	Phenyl	30	0.65	<0.5	12.5	13.3	0.19	25
Arsenic	Triphenyl	12	1.66	<1	11.5	13.0	0.19	60
Antimony	Triphenyl	15	3.3	<1	11.3	12.4	0.30	200
Germanium	Di-n-butyl	20	2.7	1.3	10.8	10.9	0.24	90
	Diphenyl	20	2.7	0.5	11.6	12.8	0.22	100
Selenium	Diphenyl	10	1.5	0.12	10.6	9.8	0.31	180

TABLE I (continued)

Metal	Organic Ligand	Mole % Metal Interm. in Feed	Polymer Analyses					Speed*
			Wt % M Calc.	Wt % M FD.	Cl Calc.	Cl FD.	{η}	
Manganese	Dipyridyl	12	1.2	0.3	11.8	12.9	0.35	1150
Iron	Ferrocene-1,1'-dicarbox.	12	1.3	0.6	10.7	10.9	0.60	600
Silicon	Dimethyl	20	1.1	1.1	11.3	11.8	0.20	3500
	Diphenyl	20	1.1	1.1	11.8	11.9	0.22	90
	Methyl phenyl	20	1.1	1.1	12.1	11.6	0.21	60
	Methyl phenyl	50					0.13	50
Palladium	Bis(triphenyl-phosphine)	4.5	0.92			11.4	0.67	
Platinum	Bis(triphenyl-phosphine)	5	3.37	<1	10.8	11.4	0.46	8500
Ruthenium	Tris(triphenyl-phosphine)	5	0.95	<0.3	11.6	11.7	0.49	7000
Titanium	Diacetyl acetonate	12	1.1	1.4	11.6	11.8	0.19	300
	Titanocene	20	1.8	2.0	11.8	12.2	0.18	40
Hafnium	Hafnocene	10	3.3	3.0	11.6	11.7	0.39	5000
Vanadium	Vanadocene	10	0.97	0.8	11.9	12.2	0.31	700
Zirconium	Zirconocene	10	1.7	1.2	11.8	12.2	0.48	7400

*
A qualitative comparison of photosensitivity in which a 2% solution of the polymer is
whirl-coated on grained aluminum and then exposed through a 0.15 density gradient step
tablet for a given time to a UV rich source, such as the Colight XPOSER printer or the
Ozalid Ozamatic printer. The sample is then solvent developed and the number of completely
cross-linked steps is noted with the corresponding density of the step tablet. This is
then compared to a similar series of coatings of poly(vinyl cinnamate) with and without
sensitizers. A sensitivity value of 2 corresponds to unsensitized poly(vinyl cinnamate)
and 1000 to poly(vinyl cinnamate) sensitized with 2-benzoylmethylene-1-methyl-β-naphth-
thiazoline. To be useful in a photoresist or photolithographic application, a 'Speed'
of at least 1000 is required. A 'Speed' of 5000 would mean that 1/5 the exposure is
required to obtain the same degree of crosslinking as with a 'Speed' of 1000.

123

TABLE II

Organometallic Polysebacates Prepared From DVCP and Tetrabromobisphenol A

Metal	Organic Ligand	Mole % Metal Interm. in Feed	Polymer Analyses				$\{\eta\}$	Speed
			Wt % M Calc.	Wt % M FD.	Wt % Br Calc.	Wt % Br FD.		
Tin	Dimethyl	25	5.1	1.7	19.2	18.8	0.27	700
	Di-n-butyl	20	3.9	0.6	18.5	18.4	0.74	15000
	Dicyclohexyl	20	3.9	<1	18.3	18.3	0.52	7000
	Diphenyl	12	2.4	0.8	18.7	18.6	0.56	6500
	Dibenzyl	25	4.8	2.9	18.1	15.6	0.18	700
	Dioctyl	20	3.8	0.5	18.0	18.0	0.57	7000
Lead	Diphenyl	12	4.0	3.4	18.4	17.8	0.60	10000
	Diphenyl	20	6.6	3.8	17.9	18.4	0.30	2300
Boron	Phenyl	20	0.4	0.3	19.6	17.9	0.34	1800
	Phenyl	35	0.7	0.6	20.0	20.9	0.14	35
Arsenic	Triphenyl	15	1.9	2.0	18.4		0.13	50
	Triphenyl	10	1.4	1.4	18.7		0.23	750
Antimony	Triphenyl	15	3.0	0.8	18.2	18.5	0.41	6000

TABLE II (continued)

Metal	Organic Ligand	Mole % Metal Interm. in Feed	Polymer Analyses				{η}	Speed
			Wt % M Calc.	Wt % M FD.	Wt % Br Calc.	Wt % Br FD.		
Germanium	Di-n-Butyl	15	1.8	2.0	19.0	18.1	0.55	7000
	Di-n-Butyl	30	3.1	0.5	18.9	20.1	0.18	60
	Di-n-Butyl	22	2.7	0.4	18.9	18.7		1150
	Dimethyl	20	2.4	0.8	18.7	18.9	0.27	600
Selenium	Diphenyl	10	2.9	<1				21000
	Diphenyl	15	2.0			23.8	0.29	600*
Manganese	Dipyridyl	12	1.1	0.5	21.5	21.8	0.38	1100
Iron	Ferrocene-1,1'-Dicarboxyl	15	1.4	1.2	18.7	15.7	0.38	200
	Ferrocene-1,1'-Dicarboxyl	12	1.1	1.6	18.8	18.6	0.53	750
Silicon	Dimethyl	25	1.8	0.8	20.0	18.8	0.17	60
	Diphenyl	25	1.2	1.2	18.9	18.8	0.18	430
	Methyl Phenyl	20	1.0	1.0	21.8	22.5	0.16	50*
Palladium	Bis(triphenyl phosphine)	4.7	0.7	2.4	18.5	18.1	0.33	1600
Platinum	Bis(triphenyl phosphine)	6	1.9		18.1	19.1	0.56	5608
Ruthenium	Tris(triphenyl phosphine)	2	0.3	0.2	18.6	17.5	0.33	4200
Titanium	Diacetyl Acetonate	15	1.3	1.4	19.2	17.4	0.46	12000
	Titanocene	12	1.0	0.4	19.1	17.9		12000
Hafnium	Hafnocene	12	3.5	5.7	20.9	22.3	0.27	3500*
Vanadium	Vanadocene	12	1.0	0.7	19.0	16.5	0.19	130
Zirconium	Zirconocene	12	1.8	1.7	21.2	24.3	0.29	1100*

* DVCP: TBrBPA 60:40 Mole Ratio - All others 65:35.

The author appreciates the work of Dr. Hans Coll, Mr. Robert Dollinger, and Miss Barbara Coulter for the GPC fractionation studies and of Mr. Gordon Meyer and Mrs. Carol Brown for their determination of the metals in the polymers.

IV.   REFERENCES

1.   Carraher, C.E., et al., J. Polym. Sci., 7, 2351 (1969); ibid., 8, 973, 3051, 3367 (1970); ibid., 9, 983, 3661 (1971); ibid., 10, 413, 524, 3367 (1972); ibid., 12, 799 (1975); ibid., 13, 215 (1975).

2.   Carraher, C.E., et al., Die Makromolekulare Chemie, 130, 177 (1969); ibid., 131, 259 (1970); ibid., 133, 211 (1970); ibid., 135, 107 (1970); ibid., 141, 237, 259 (1971); ibid., 43, 49, 61 (1972); ibid., 160, 251 (1972); ibid., 164, 87 (1973); ibid., 166, 23, 31 (1973); ibid., 175, 2307, 3089 (1974).

3.   Carraher, C.E., et al., Polymer Preprints, 10, 418 (1969); ibid., 11, 66, 606 (1970).

4.   Carraher, C.E., et al., Polymer, 13, 153 (1972); ibid., 14, 42 (1973); ibid., 15, 9 (1974).

5.   Carraher, C.E., et al., Macromolecules, 2(3), 306 (1969); ibid., 4(2), 264 (1971); C.U. Pittman, G.V. Marlin, T.D. Rounsefell, ibid., 6(1), 1 (1973).

6.   Carraher, C.E., Chemtech, 741 (Dec. 1972).

7.   Carraher, C.E., et al., Die Angewandte Makromolekulare Chemie, 28, 145 (1973); ibid., 25, 121 (1972); ibid., 28, 153 (1973); ibid., 115 (1973); ibid., 57 (1974); ibid., 39, 69 (1974).

8.   Carraher, C.E., et al., J. Appl. Polym. Sci., 16(7), 1851 (1972); ibid., 18, 53 (1974).

9.   Carraher, C.E., et al., Brit. Polym. J., 6, 255 (1974); ibid., 7, 155 (1975).

10. Carraher, C.E., and Lessek, P., Eur. Polym. J., 8, 1339 (1972).

11. Carraher, C.E., et al., J. Macromol. Sci. Chem., 7, 913 (1973); 7, 1349 (1973); ibid., A8, 1009, 1249 (1974); ibid., A9, 191 (1975).

12. Carraher, C.E., et al., Am. Chem. Soc. Div. Org. Coat. Plast. Chem. Papers 1973, 33(1), 629; ibid., (2), 2, 427 (1973).

13. Neuse, E.W., and Rosenberg, H., "Metallocene Polymers," Reviews in Macromolecular Chemistry, Marcel Dekker, New York, 1970.

14. Almond, L., Organometallic Compounds, 27(6), Sept. 1975, 156-166, "Ferrocene Polymers, Their Preparation and Applications, 1961-1975."

15. Borden, D.G., Unruh, C.C., and Merrill, S.H., U.S. Pats. 3,453,237, 3,647,444 (1969).

16.   Arcesi, J.A., and Rauner, F.J., U.S. Pat. 3,640,722
      (1972).
17.   Arcesi, J.A., Rauner, F.J., and Williams, J.L.R., J. Appl.
      Polym. Sci., 15, 513 (1971).
18.   Borden, D.G., Polym. Eng. Sci., 14(7), 487 (1974).
19.   Borden, D.G., J. Appl. Polym. Sci., in press.
20.   Borden, D.G., Research Disclosure, 143, No. 14339
      (March 1976).

ORGANOMETALLIC POLYMERS AS CATALYSTS

Robert H. Grubbs, Shiu-Chin H. Su
Michigan State University

Organometallic polymers have been used as catalysts for a large number of reactions. These catalysts are prepared by attaching a homogeneous catalyst to a polymeric support by a linking ligand. In order to understand some of the aspects of these catalysts, two analytical techniques have been explored. Election microprobe spectroscopy has been used to measure the radial distribution of the catalyst and linking agent inside of a support. Phosphorus nuclear magnetic resonance spectroscopy has been found to be an efficient, rapid means of analysis of low cross-linked phosphinated polymers. The results give a structure for the catalyst on the polymer.

I. INTRODUCTION

Organometallic polymers show many potential uses as structural material with desirable chemical and physical properties. Another application of this class of polymers is in catalysis. (1) Most of the polymers prepared for this usage are prepared by the linking of a homogeneous catalyst to a preformed polymer. The polymer is functionalized to provide a ligand for linking the catalyst to the polymer. A majority of the systems to date have used cross-linked polystyrene as an organic or silica gel as an inorganic support. In the following discussion only the organic systems will be considered.

It has been demonstrated that almost any homogeneous catalyst can be attached to an organic support by replacing one or more of the ligands on the soluble catalyst with an insoluble polymeric ligand. The complex retains its basic catalytic properties after attachment. Superimposed on the inherent properties of the catalyst are changes in activity and selectivity resulting from support-catalyst interactions. The attachment usually results in a decrease in activity[1],

except in special cases[2], a change in selectivity of a substrate on the basis of size[3] and polarity, and a change in
regioselectivity[4]. Many of these changes are a function of
loading of the support[5]. One of the main advantages of
these catalysts is the ease of separation of the catalyst in a
reusable form from the reaction mixture[6].

II.   POLYMER PREPARATION

Numerous techniques have been developed for the function-
alization of polystyrene with ligands appropriate for metal
attachment. The most used are phosphine, amine, and cyclopentadienyl ligands.

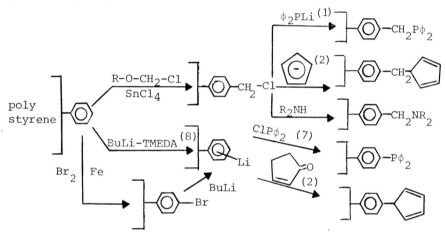

Examples of attachment of metals to these polymeric ligands
are:

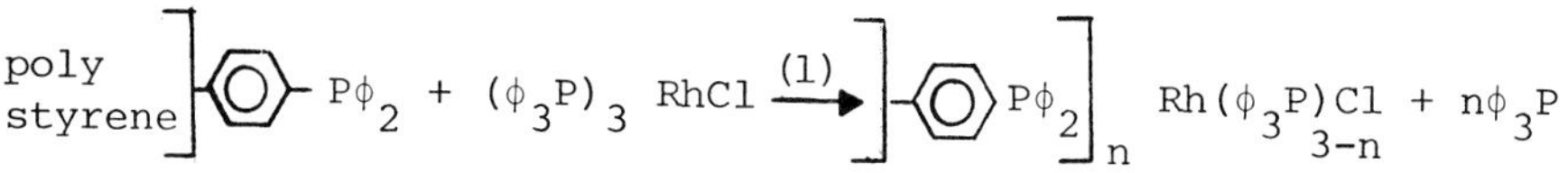

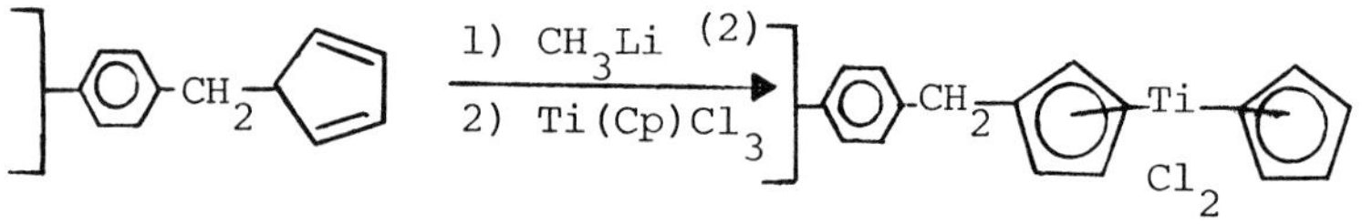

The understanding of the factors which control the changes
in catalytic activity on polymer attachment have been difficult
to determine due to the lack of good methods for the analysis

of the structure of the catalyst on the polymer. Two
techniques we have found useful for the analysis of the
changes in catalytic activity on polymer attachment are electron microprobe spectroscopy[9] and nuclear magnetic resonance
spectroscopy.

III.   ELECTRON MICROPROBE

Electron microprobe analysis allows the rapid determination of the distribution of the catalyst inside of a polymer
bead. This is particularly important when studying reactions
which are a function of the loading of the ligand and the
catalyst on polymer. Loading levels are generally determined
from bulk elemental analysis of the catalyst. If either the
catalyst or the linking ligand are not homogeneously distributed throughout the insoluble polymeric support, the local
loading levels will be different than that determined by
analysis of the polymer bead. Many of the substitution reactions given earlier do not result in homogeneous substitution
of the polymer. Consequently, it is essential to determine
the local loading of both ligand and catalyst. For microprobe
analysis, a polymer bead is halved and the resulting cross-
section is mounted in the apparatus. An electron beam is
passed across the surface and the x-ray emission as a function
of distance across the center of the bead is measured. By the
proper turning of the detector, the intensity of the characteristic emission of each element in the surface can be measured as a function of the distance into the bead. By
measuring the relative radial distribution of the ligand
element and the catalyst metal the local concentration of each
can be determined

IV.   SPECTROSCOPY

Catalyst attached to polystyrene by phosphine links have
received the most study (1,2,3). For example:

$$\text{poly styrene} - \left[\bigcirc - P \right] + \left[(\text{cyclooctene})_2 ClRh \right]_2 \longrightarrow \left[\bigcirc - P - \right]_n Rh\text{-}Cl \qquad 1)$$

The relationship of structure to activity and selectivity
have been examined in the most detail for both homogeneous and
hybrid rhodium (I) phosphine hydrogenation [1], and hydrofor-

mylation catalysts (1,5). In both of these cases the structure, as well as its selectivity and activity, of the homogeneous catalyst changes with changes in the metal to phosphine ratio. In order to understand the activity and selectivity changes of the hybrid catalysts as a function of conditions used in the preparation a rapid, simple probe of catalyst structure was required.

^{31}p nmr spectroscopy was chosen to study these phosphine catalyst systems since this nucleus is easily studied at low concentration, the ^{31}p spectra of a number of phosphine metal (10) complexes have been examined, and the ^{31}p chemical shift is very sensitive to structural changes. The ^{31}p chemical shift of phosphinated polystyrene (prepared from 2%-divinyl-benzene-styrene copolymer) swollen in toluene is identical to that of triphenylphosphine. The chemical shift difference between the phosphine and phosphineoxide is large. Consequent-ly, the amount of oxidized phosphine on the polymer is easily determined. When a metal complex ($[(cyclooctene)_2 ClRh]_2$ eq. 1) was added to the nmr sample, the intensity of the ^{31}p signal decreased relative to an external standard and no other peaks appeared. This apparently results from an increase in the relaxation rate on complexation of the phosphine to a metal. The portion of the ^{31}p signal intensity which disappears on the addition of the metal is a measure of the percentage of the phosphine complexed. If the number of mmoles of phosphine on the starting polymer and the mmole of rhodium is easily calculated. Figure 1 shows a plot of the amount of phosphine complexed in a polymer sample containing 0.72 mmoles P/gm beads as a function of the amount of metal added. As can be seen, the amount of phosphine complexed (Pc) increases linearly until the polymer becomes saturated. Table 1 gives the ratio of the mmoles of phosphine complexed (Pc) to the mmoles of rhodium added.

TABLE I

Sample	meq Rh(added)/g	meq Pc/g	Meq Pc/meq Rh
A	0.04598	0.1579	3.43/1
C	0.07717	0.2137	2.77/1
D	0.10632	0.2799	2.63/1
E	0.20664	0.4486	2.17/1
F	exs	0.6251	1.82/1

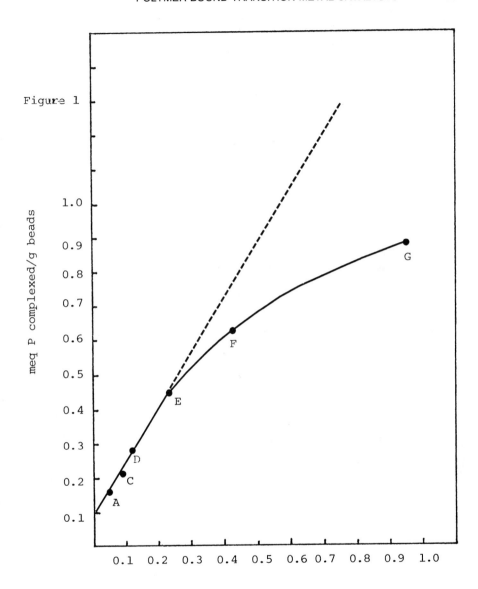

Figure 1

Figure 1:   Meq of phosphorus complexed as a function of rhodium
            complex added per meq of total phosphines (p).

As can be seen, the structure of the complex slowly changes from a trisphosphine complex as the predominate form to a bisphosphine system at high loading. The intermediate loadings should contain a mixture of these two extreme complexes. This technique then provides a rapid method for the determination of the molecular structure of complexes attached to low cross-linked swellable polymers. The information obtained by this method can now be used to explain the rate/loading relationship of a number of catalysts.

V. REFERENCES

1a. E.M. Cernia and M. Graziani, J. Appl. Polymer Sci., 18, 2725 (1974).

b. C.U. Pittman and G.O. Evans, Chem. Tech., 560 (1973).

c. J.C. Bailar, Jr., Cat. Rev. - Sci. Eng., 10, 197 (1974).

d. J. Manassen, J. Chim. Ind., 51 (1969).

e. W.O. Haag, et al., Belgian Patent. 721,686 (1968).

f. H.O. Haag and D.D. Whitehurst, Second North American Meeting of the Catalyst Society, 1971, Houston, Texas.

g. R.H. Grubbs and L.C. Kroll, J. Am. Chem. Soc., 93, 3062 3062 (1971).

2. W.D. Bonds, Jr., C.H. Brubaker, Jr., E.S. Chandrasekaran, C. Gibbons, R.H. Grubbs, and L.C. Kroll, J. Am. Chem. Chem Soc., 97, 2128 (1975).

3. R.H. Grubbs, L.C. Kroll, and E.M. Sweet, J. Macromol. Sci.-Chem., A7, 1047 (1973).

4. R.H. Grubbs, E.M. Sweet, and S. Phisanbut, Catalysis in Organic Synthesis, Acad. Press, Inc., 153, 1976.

5. C.U. Pittman, Jr. and R.M. Hanes, J. Am. Chem. Soc., 98, 5402 (1976).

6. K.G. Allum, P.D. Hancock, I.V. Howell, R.C. Pitkethly, and P.J. Robinson, J. Organomet. Chem., 87, 189 (1975) and references therein.

7. J.P. Collman, L.S. Hegedus, M.P. Cooke, J.R. Norton, G. Dolcetti, and D.N. Marquart, J. Am. Chem. Soc., 94, 1748 (1972).

8. T.M. Fylesand and C.C. Leznoff, Can. J. Chem., 54, 935 (1976); and R.H. Grubbs and S.C. H. Su, J. Organomet. Chem., 122, 151 (1976).

9. R.H. Grubbs and E.M. Sweet, Macromocules, 8, 241 (1975).

10. M.M. Crutchfield, C.H. Dungan, J.H. Letcher, V. Mark and J.R. VanWazer, Topics in Phosphorous Chemistry, Inter- science Pub., New York, 1967.

SUPPORTED CATALYSTS FOR ETHYLENE POLYMERIZATION

Frederick J. Karol
Union Carbide Corporation

Supported catalysts for ethylene polymerization may be divided into three classes: (1) metal oxide, particularly CrO_3/SiO_2; (2) Ziegler, particularly $R_3Al+TiCl_4/Mg(OH)Cl$; (3) organotransition metal compounds, particularly $(C_5H_5)_2Cr/SiO_2$. A significant improvement in polymerization activity probably represents the single most important advantage of supported catalysts. These supported catalyst systems show high efficiency presumably because the active transition metal compound resides only on the support surface, permitting availability of a larger concentration of active sites for polymerization. Reaction of a transition metal compound with a support surface provides an anchoring device, preventing destruction of potential sites by mutual interaction. Studies with the CrO_3/SiO_2 catalysts have shown that formation of a surface chromate takes place by reaction of CrO_3 and surface silanol groups on silica. Polymer chain growth is believed to occur by a coordinated anionic mechanism. With the Ziegler catalysts, $R_3Al+TiCl_4$ supported on magnesium hydroxychloride, chemisorption can be represented by: $TiCl_4+Mg(OH)Cl \rightarrow Cl_3TiO-MgCl+HCl$. After reduction by aluminum alkyl, formation of a highly active catalyst occurs. Chromocene deposited on silica supports forms a highly active catalyst for ethylene polymerization. The catalyst formation step liberates cyclopentadiene and leads to a new, chemically-anchored chromium species attached to one cyclopentadienyl ligand. This catalyst shows a high response to hydrogen. Exchange of ligands at the active sites of these organotransition metal catalysts represents a potentially powerful tool for regulating catalytic behavior.

I. INTRODUCTION

Supported transition metal catalysts are widely used for the production of high density, linear polyethylene and ethylene-α-olefin copolymers. These supported catalysts have attracted considerable attention and appear to be an object of detailed study in many industrial and academic laboratories throughout the world. Numerous publications,[1-3] particularly in the patent literature and trade journals, describe new developments and discuss current and future activities in this exciting area of catalysis. The present paper focuses on some chemistry of catalyst formation for the different classes of supported catalysts. It also deals with the effect of ligand environment at the active site on polymerization behavior.

II. DISCUSSION

Supported transition metal catalysts for ethylene polymerization may be divided into three broad classes:

1) Metal oxide, particularly CrO_3/SiO_2
2) Ziegler-Natta, particularly $R_3Al+TiCl_4/Mg(OH)Cl$
3) Organotransition metal compounds, particularly $(C_5H_5)_2Cr/SiO_2$

This classification of types of supported catalysts attempts to acknowledge systems that have received serious attention and offer commercial status or potential, and to recognize certain other systems that illustrate the scope of investigations with supported catalysts. Catalyst productivities are so high ($\sim 1 \times 10^5 - 1 \times 10^6$ g polymer/g transition metal) that, in many cases, it is not necessary to remove catalyst residues after polymerization. Commercial routes for production of linear, high density polyethylene include solution, slurry, and gas phase processes. In each process appropriate control of catalyst preparation conditions and reactor parameters permit production of a wide range of polyethylenes to satisfy specific end-use purposes.

A. Metal Oxide Catalysts

Chromium oxide catalysts, supported on silica or silica-alumina, represent the most important metal oxide for ethylene polymerization. A significant portion of current worldwide production of high density polyethylene is made with this catalyst.

The chromium oxide polymerization catalysts can be readily made by impregnation of high surface area silica-alumina or silica with an aqueous solution of a soluble chromium compound such as chromium trioxide (CrO_3).[4,5] The finely divided catalyst, after being dried, is normally activated by fluidizing in air at 500-1000°C for several hours.

Chromium trioxide reacts with silica to form a new surface compound. This new compound has higher thermal stability than chromium trioxide itself. Formation of a surface chromate by reaction with CrO_3 and silanol groups is believed to occur in the manner shown by (eq. 1).

$$\begin{array}{c} \text{-Si-OH} \\ \text{O} \\ \text{-Si-OH} \end{array} + CrO_3 \rightleftharpoons \begin{array}{c} \text{-Si-O} \\ \text{O} \\ \text{-Si-O} \end{array}\!\!Cr\!\!\begin{array}{c} \text{O} \\ \text{O} \end{array} + H_2O \qquad (1)$$

An increase in activation temperature, with removal of the water formed during chemisorption, will increase the concentration of surface chromate. Addition of an excess of water leads to a shift to unreacted, bulk CrO_3, which will decompose to Cr_2O_3 and O_2. The necessity of using oxygen or air during the heating cycle prevents rapid decomposition of hexavalent chromium. A highly active catalyst requires a delicate balance between removal of water and maintaining chromium in its highest valence state.

Reaction of the chemisorbed chromate with ethylene results in an oxidation-reduction process with formation of a low-valent chromium center (eq. 2).[6,7]

$$\begin{array}{c} \text{-Si-O} \\ \text{O} \\ \text{-Si-O} \end{array}\!\!Cr\!\!\begin{array}{c} \text{O} \\ \text{O} \end{array} + CH_2{=}CH_2 \longrightarrow \begin{array}{c} \text{-Si-O} \\ \text{O} \\ \text{-Si-O} \end{array}\!\!Cr + \text{oxidation products}$$
$$(CO, H_2) \qquad\qquad (CO_2, H_2O)$$
$$(2)$$

The primary initiation reaction remains unclear, but is believed to result in formation of a chromium hydride or alkyl. Polymer chain growth may be pictured to occur by a coordinated anionic mechanism (eq. 3).[4,8]

$$Cr{-}R + nCH_2{=}CH_2 \longrightarrow Cr(CH_2{-}CH_2)_n R \text{ etc.} \qquad (3)$$
$$R = \text{hydride or alkyl}$$

Termination of the growing polymer chain can occur by hydride-ion transfer to monomer or catalyst. Each polyethylene molecule has a methyl group at one end and a vinyl group at the other end of the polymer chain.

At a 30-100 ppm chromium loading, chromium efficiency was at a maximum.[4] Each chromium atom at this loading was assumed to be an active site. The concentration of active sites was calculated to be 2.5×10^{-5} mole/g of catalyst. The time required for termination of a growing chain and initiation of a new one was determined to be quite small. Calculations showed that the addition rate was 2000 molecules/sec. and 2.8 polymer molecules were produced per second per active site. During the polymerization process catalyst particles do not necessarily remain in the middle of a polymer particle but shatter and are scattered throughout the polymer.

B. Supported Ziegler-Natta Catalysts

High activity Ziegler-Natta catalysts, based on reaction products of specific magnesium, titanium, and aluminum compounds, have been developed for ethylene and propylene polymerization. Catalysts, chemically anchored on Mg(OH)Cl-type supports, were pioneered by Solvay and provided much of the early impetus in this area.[9,10] The essential feature of these catalysts is that the transition metal compound is chemically attached to the surface of a solid magnesium compound. It is believed that chemisorption takes place by reaction of a halogen atom from the transition metal compound with hydroxyl groups at the surface of the magnesium compound, which results in transition metal-oxygen bonds (eq. 4).

$$\text{TiCl}_4 + \text{Mg}\Big\langle {}^{\text{Cl}}_{\text{OH}} \longrightarrow \text{Mg}\Big\langle {}^{\text{Cl}}_{\text{OTiCl}_3} + \text{HCl} \qquad (4)$$

Some characteristic supported magnesium-titanium catalyst components include:

$$\begin{aligned}
&\text{Mg(OH)Cl} + \text{TiCl}_4 \\
&\text{MgCl}_2 \cdot 3\ \text{Mg(OH)}_2 + \text{Ti(OR)}_x\text{Cl}_y \\
&\text{MgSO}_4 \cdot 3\ \text{Mg(OH)}_2 + \text{Ti(OR)}_x\text{Cl}_y \\
&\text{Mg(OH)}_2 + \text{TiCl}_4
\end{aligned}$$

These supported titanium compounds are activated by addition of aluminum alkyls to provide active catalysts with structures of the type illustrated by (eq. 5).

$$\text{Mg}\Big\langle {}^{\text{Cl}}_{\text{OTiCl}_3} + \text{R}_3\text{Al} \longrightarrow \text{Mg}\Big\langle {}^{\text{Cl}}_{\text{O—Ti}} \cdots \text{Al} \qquad (5)$$

Other high activity Ziegler-Natta catalysts may be prepared from the reaction products of magnesium alkoxides with transition metal halides or the reaction products of magnesium

chloride with transition metal compounds.[2,3] Chemical anchoring with these catalysts appears to occur through chlorine
bridging with the magnesium substrate.

Supported Ziegler-Natta catalysts are claimed to exhibit
sufficiently high activity that catalyst residues do not have
to be removed from the polymer. Catalyst parameters which may
be used to introduce process and product flexibility include
the following:[2]

> Nature and valency of the transition metal.
> Ligand environment at the transition metal.
> Nature of the organometallic reducing agent.
> Catalyst morphology including crystallinity, porosity,
> external and internal surface area.

Significant progress has been made towards higher effi-
ciencies and selectivity with the high activity Ziegler-Natta
catalysts.[2] The amount of polyethylene produced per titanium has been increased more than twentyfold. Polyethylenes of
either narrow ($\bar{M}_w/\bar{M}_n$ = 4) or broad ($\bar{M}_w/\bar{M}_n$ = 24) molecular
weight distribution are claimed to be possible by changing the
composition of a single catalyst type. Polymer molecular
weight can be controlled by addition of hydrogen as a chain
transfer agent.

C. Organotransition Metal Compounds

The direct use of organometallic compounds of transition
metals for the preparation of solid catalysts for olefin polymerization developed in the 1960's. Catalysts obtained by
supporting π -cyclopentadienyl, π-allyl, and σ-organometallic
compounds of certain transition metals proved to be active for
ethylene polymerization.[11-13] The interaction between organometallic compounds and oxide supports for the preparation
of catalysts can be illustrated by the following schemes. Reactions of the type illustrated by eqs. (6) and (7) are
possible for the interaction of organometallic compounds
MR_yL_x or MR_z with surface hydroxyl groups of the support.

$$(E-OH)_n + MR_yL_x \longrightarrow (E-O)_n-ML_xR_y-n + n\ RH \qquad (6)$$

$$(E-OH)_n + MR_z \longrightarrow (E-O)_nMR_{z-n} + n\ RH \qquad (7)$$

$$M = \text{transition element}$$
$$R = \text{organic ligand}$$
$$L = \text{anion or inorganic ligand}$$
$$E = \text{element forming oxide support}$$

The composition of surface compounds in these catalysts is influenced by the temperature of preliminary dehydration of the support which determines the number of surface reactive groups.

Chromocene deposited on silica of high surface area forms a highly active catalyst for polymerization of ethylene. The catalyst formation step (eq. 8) liberates cyclopentadiene and leads to a new divalent chromium species containing a cyclopentadienyl ligand.

$$(8)$$

Polymerization is believed to occur by a coordinated anionic mechanism (eq. 9). The catalyst has a very high chain transfer response to hydrogen (eq. 10) which permits facile preparation of a full range of polymer molecular weights.[14,15]

$$(9)$$

$$(10)$$

Catalyst activity increases with an increase in silica dehydration temperature, chromium content on silica, and ethylene reaction pressure. Oxygen addition studies show activity is proportional to initial divalent chromium content.

Silica-supported bis (indenyl)-chromium, $(C_9H_7)_2Cr/SiO_2$, and bis (fluorenyl)-chromium, $(C_{13}H_9)_2Cr/SiO_2$, also show good activity in ethylene polymerization.[16] The polymerization activity of the supported indenyl and fluorenyl chromium catalysts illustrates that large π-organic ligands initially attached to chromium will not totally retard the ethylene polymerization process. Sufficient space must be available between the support surface and the "umbrella" of the organic ligand for ethylene to coordinate and insert at the active chromium site. By analogy with the results obtained with the supported chromocene catalyst, the active sites are believed to contain an indenyl or fluorenyl ligand attached to a divalent chromium center:

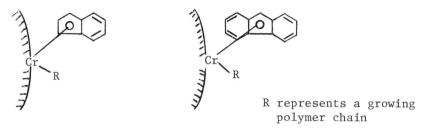

R represents a growing
polymer chain

Addition of cyclopentadiene to a supported indenyl chromium catalyst provided a catalyst with a much higher transfer
response to hydrogen. The increase in hydrogen response of the
supported indenyl chromium catalyst suggests that a ligand exchange (eq. 11) can occur.

$$\text{(11)}$$

Thermal aging of a chromocene catalyst in an inert atmosphere leads to a modified catalyst which shows poor response to hydrogen as a transfer agent.[17] By thermogravimetry the weight loss of the catalyst, relative to dehydrated
silica, was equivalent to loss of one cyclopentadienyl ligand
per chromium site. Pyrolytic gas chromatography showed cyclopentadiene was liberated in the thermal process (eq. 12).
These studies provide strong evidence that loss of a cyclopentadienyl ligand in supported chromium catalysts has a profound effect on overall polymerization behavior.

$$\text{+ organic fragments} \qquad \text{(12)}$$

Highly active catalysts for ethylene polymerization are also
formed when bis (benzene)- or bis (cumene)-chromium or tris
and bis (allyl)-chromium compounds are deposited on high surface area silica-alumina or silica supports.[18] Each catalyst type shows its own unique behavior in preparation, polymerization activity, and response to hydrogen as a chain
transfer agent. Arene chromium catalysts produced high molecular weight polyethylene and showed, in contrast to supported chromocene catalysts, a much lower response to hydrogen

as a chain transfer agent. Addition of cyclopentadiene to supported bis (cumene)-chromium catalyst led to a new catalyst which showed chain transfer response to hydrogen typical of a supported chromocene catalyst (eq. 13).

$$\text{Bis (cumene)-chromium} + \text{\Large\ding{}} \longrightarrow \text{chromocene-type catalyst} \quad\quad (13)$$
catalyst

Polymerization activity with tris or bis (allyl)-chromium catalyst appears to depend on the divalent chromium content of the catalyst (eqs. 14, 15).

$$\text{Tris (allyl)-chromium} \longrightarrow \text{Bis (allyl) chromium (II)} \quad (14)$$

Bis (allyl)-chromium (II) + $\left.\right\}$Si$\underset{OH}{\langle}$ $\longrightarrow$ supported allyl chromium catalyst + propylene $\quad (15)$

The order of activity of the different types of catalysts was chromocene/silica > chromocene/silica-alumina > bis (arene)-chromium/silica-alumina $\cong$ allyl chromium/silica.

D.   General Features of Supported Catalysts

The current status of ethylene polymerization catalyzed by supported transition metal compounds, which function by a coordinated anionic mechanism, suggests a number of general conclusions. Catalyst supports such as silica can act as a chemical reactants to generate active sites (or precursors). The three types of supported catalysts illustrate such a role for the substrate.

Several routes to the active site or initial transition metal-carbon bond exist. Transition metal-carbon bonds may be formed by alkylation of a supported transition metal compound using a metal alkyl (Ziegler). Reduction by ethylene which occurs with the CrO_3/SiO_2 catalyst provides another, different route in the process for formation of the active site. Finally, displacement of an organic ligand by the substrate to provide new, supported transition metal compounds with available coordination positions, also furnishes a route to active sites.

Ligand environment at the active site plays a significant role in determining response parameters of the catalysts. Exchange of ligands at these sites represents a potentially powerful tool for regulating catalytic behavior. The behavior and characteristics of these transition metal species, chemically anchored to a substrate, can be expected to receive considerable attention in the future.

III. REFERENCES

1. Karol, F. J., Encyclopedia of Polymer Science and Technology, Supplement I, pp 120–146 (1976) and references therein.
2. Weissermel, K., Cherdron, H., Berthold, J., Diedrich, B., Keil, K. D., Rust, K., Strametz, H., and Toth, T., J. Polym. Sci., Symp. 51, 187 (1975) and references therein.
3. Diedrich, B., Appl. Polym. Symp., 26, 1 (1975) and references therein.
4. Hogan, J.P., J. Polym. Sci., A-1, 8, 2637 (1970).
5. Hogan, J. P., and Banks, R. L., (to Phillips Petroleum Co.) U.S. Pat. 2,825,721 (1958).
6. Krauss, H. L. and Stach, H., Z. Anorg. Allg. Chem., 366, 280 (1969).
7. Baker, L. M., and Carrick, W. L., J. Org. Chem., 33, 616 (1968).
8. Karol, F. J., in "Applied Polymer Science" (J. Cramer and R. Tess, Eds.) Chapter 15, ACS Publication, 1975.
9. Stevens, J., Hydrocarbon Process, November 1970, pp 179–182.
10. Dassesse, P., and Dechenne, R., (to Solvay and Cie, Belgium), U.S. Pat. 3,400,110 (1968).
11. Karol, F. J., Karapinka, G. L., Wu, C., Dow, A. W., Johnson, R. N., and Carrick, W. L., J. Polym. Sci., A-1 10, 2621 (1972).
12. Yermakov, Y. I., Catal. Rev.-Sci. Eng., 13 (1), 77 (1976).
13. Ballard, D. G. H., Advan. Catal., 23, 263 (1973).
14. Karol, F. J., Brown, G. L., and Davison, J. M., J. Polym. Sci., Polym. Chem. Ed., 11, 413 (1973).
15. Karapinka, G.L., (to Union Carbide Corp.) U.S. Pat. 3,709,853 (1973).
16. Karol, F.J., (to Union Carbide Corp.) U.S. Pat. 4,015,059 (1977).
17. Karol, F. J., and Wu, C., J. Polym. Sci., Polym. Chem. Ed., 12, 1549 (1974).
18. Karol, F. J., and Johnson, R. N., J. Polym. Sci., Polym. Chem. Ed., 13, 1607 (1975).

RHODIUM ELUTION FROM POLYMER-BONDED HYDROFORMYLATION CATALYSTS

W. H. Lang,* A. T. Jurewicz,**W. O. Haag
D. D. Whitehurst and L. D. Rollmann

Mobil Research and Development Corporation
Princeton Laboratory

ABSTRACT

The concentration of rhodium in solution over polymer-bonded hydroformylation catalysts was measured, and the factors influencing that concentration were identified. Rhodium dissolution exhibited an equilibrium behavior which decreased with increasing temperature or hydrogen pressure and with decreasing CO pressure. Solvent effects were particularly pronounced with amine resins and the sequential reaction, olefin-aldehyde-alcohol, led to significant redistribution of rhodium within a reactor bed. The implications of these effects in catalysis are noted.

I.   Introduction

In recent years understanding of homogeneous catalysis by transition metal complexes has been characterized by an extraordinary detail. It is now well recognized that potentially advantageous effects can be realized if ligands are made a part of an insoluble but accessible "catalyst support," supports including polymers, resins, and silicas, among others (1-6). In addition to simply facilitating or eliminating catalyst recovery operations,

---

*   To whom correspondence should be addressed.

**   Mobil Chemical, Edison, New Jersey 08817

matrix bonding can afford materials of unusual
selectivity, such as a rhodium-based catalyst
for the one-step synthesis of OXO alcohols (2)
or a selective olefin hydrogenation system (6).
It can also stabilize unusual chemical species
such as titanocene (5).

Implicit in the use of matrix-bonding techniques
is the assumption that the metal catalyst complex
remains intact throughout the catalytic cycle,
i.e., that the metal ion remains attached (coordinated) to the insoluble ligand. Supporting this
assumption are long-term use in continuous-flow
systems (1), but only limited information exists
on the actual rhodium concentration in equilibrium
over these catalysts (4,8). In the following
pages data are presented which detail the
equilibria involved in the elution of rhodium from
one such reacting system, a macro-reticular
phosphine resin, and which show the influence of
process variables on those equilibria.

II.   Experimental

A.   Materials

Resin-starting materials were
obtained from Rohm and Haas and from Ionac. All
were porous, macroreticular, styrene-based
polymers, cross-linked with divinylbenzene or
with ethylene glycol dimethacrylate. Amine
resins were polymeric analogs of N,N-dimethyl-
benzylamine and contained 5.8% nitrogen (4.1
meq per gram). Phosphine resins were analogs of
dibutylphenylphosphine. These latter resins
contained 6.6-9.7% phosphorus (2-3 meq per
gram).

Catalyst complexes were prepared by a bridge-
splitting reaction between the polymeric ligand
and rhodium carbonyl chloride (in hexane or
benzene) and initially contained the species,
cis-$Rh(CO)_2$(polymer)Cl, as evidenced by infrared (8). Rhodium levels never exceeded 2%
(0.2 meq/g) such that the catalysts contained
at least a ten-fold excess of ligand.

B.   Procedures

Catalysts were tested in a down-flow, stainless steel reactor, heated by a circulating oil bath (50-120°C, 500-2000 psig). The reactor bed measured 30-50 cm long, 1.2 cm diameter, with a 0.3 cm thermowell in the center, and was preceded by a 20 cm preheater. Provision was made at the preheater inlet for hydrogen, carbon monoxide, and liquid feed. Reactors operated continuously, 24 hours a day.

Liquid samples (and recovered catalysts) were analyzed for rhodium spectrophotometrically, using a stannous chloride method adapted from Sandell (9). Practical limits of the measurements were 0.04 ppm Rh. Reaction products were analyzed by gas chromatography using tricresylphosphate and Carbowax 1000 columns. The olefins, 1-hexene and 1-octene (Humphrey), were percolated over alumina prior to use.

III. Results

The reactions effected by these rhodium catalysts in the presence of CO and H_2 include the isomerization of 1-hexene to internal olefins, hydroformylation to linear and branched aldehydes and, in some cases, subsequent hydrogenation of aldehyde to alcohol. Under the mild conditions of these experiments, characteristics of the various catalysts have been published and can be summarized as follows (1-2): [a] Except in the presence of certain tertiary amine ligands (pk_b = 3-7), aldehyde formation dominates with little further hydrogenation to alcohol; [b] Olefin isomerization is inhibited by phosphine to yield products high in linear aldehyde; and [c] Soluble $Rh_2(CO)_4Cl_2$ alone (or Rh carbonyl clusters) produce aldehyde and internal olefin. Indeed the high aldehyde linearity obtained over phosphine resin catalysts provides the strongest evidence that rhodium is attached to such polymers during reaction (1). Direct measurement of the polymer-solution equilibrium is detailed below.

1. Phosphine Resins

An orientation to the transport properties of rhodium on phosphine resins was provided
by the metal distribution in recovered, sectioned
reactor beds following extended processing of
olefins. Rhodium depletion was always greatest at
the top of the bed, where the resin-solution
equilibrium was established, and decreased down the
reactor.

As shown in Table I, rhodium concentrations in
benzene and in ethylhexanal solution over
phosphine resin catalysts were approximately proportional to percent loading, indicating that equilibrium is indeed established in these two solvents.
Since the phosphine ligand was in at least a 20-
fold excess over rhodium in these examples, the
ratio of rhodium concentration in solution/
rhodium concentration in resin (bound) is a good
approximation to an equilibrium constant. For
convenience the resin concentration is expressed
as weight percent loading.

The equilibrium values reveal that rhodium concentrations are strongly dependent on solvent. Where
olefins are involved equilibrium is also dependent
on conversion. As solvent polarity or as olefin
content of the liquid increased, rhodium concentration in solution increased. At 20% conversion
of a pure 1-hexene feed (1500 psig 1:1 H_2:CO,85°C),
a ratio of 6.6 x 10^{-5} was observed. Under practical
conditions of high (80-95%) conversion, solution
concentrations were described by the ratios,
1.3 x 10^{-5} (1-hexene) and 3.5 x 10^{-5} moles per
liter/percent loading (propylene), both determined
at 85°C, 1500 psig 1:1 H_2:CO.

TABLE I

POLYMER-SOLUTION EQUILIBRIA FOR
RHODIUM OVER PHOSPHINE RESINS
1000 psig H_2, 1000 psig CO, 100°C

Feed	Load-ing	[Rhodium]	Ratio[a]
Benzene	0.08%	0.15 x 10^{-5}	1.9 x 10^{-5}
Benzene	0.64%	0.8 x 10^{-5}M	1.2 x 10^{-5}

TABLE 1 (Continued

Feed	Loading	[Rhodium]	Ratio[a]
2-Ethyl-	0.08%	0.4×10^{-5}	5.0×10^{-5}
Hexanal	0.64%	2.3×10^{-5}	3.6×10^{-5}

a-- [Rhodium[/Loading

Pressure and temperature have an equally pronounced effect on rhodium elution. In an only partially successful attempt to minimize the effects of solvent and of conversion, rhodium concentrations were monitored with a 10% 1-octene, 10% methanol in hexane feed as shown in Table II. The data show three significant trends: [a] As expected, an increase in CO pressure effects a marked increase in rhodium concentration in solution; [b] Rhodium elution is sharply reduced by an increase in temperature; and [c] At higher temperatures, an increase in H_2 pressure retards the loss of rhodium.

TABLE II

PRESSURE AND TEMPERATURE EFFECTS
ON RHODIUM EQUILIBRIA
10% 1-octene, 10% methanol in hexane
At High Temperature and Conversion

Temp.	CO	H_2	[Rh]/Loading
120°CC	200	750	1.3×10^{-5}
	750	750	7.4×10^{-5}
	250	250	5.7×10^{-5}
	250	1000	2.0×10^{-5}
85°	250	1000	5.0×10^{-5}

At Low Temperature and Conversion

Temp.	CO	H_2	[Rh]/Loading
50°	200	750	12×10^{-5}
	750	750	39×10^{-5}
	250	250	22×10^{-5}
	250	1000	20×10^{-5}

The effects of CO and of temperature can be readily understood in terms of a shift in equilibrium from phosphine to carbonyl complex, increased temperature favoring the more stable phosphine. The effect of hydrogen, particularly pronounced at the highest temperature, is similarly rationalized, the hydrido phosphine complex being well known and stable in contrast to the hydrido carbonyl. There was no evidence of rhodium metal deposition in any of these experiments.

2. Amine Resins

With the amine resin catalysts, rhodium was redistributed during use such that actual increases in metal content occurred in the middle or lower sections of a reactor bed.

Since conversion over the amine resin catalyst produces first aldehyde, then alcohol, rhodium equilibria were determined first at low conversion, where aldehyde was the overwhelming product. These data are shown in Table III across a 10-fold range in Rh loading with effects of conversion shown at the bottom. As indicated, equilibrium ratios are $\sim 7 \times 10^{-5}$ at low conversion (where aldehyde is the product) and decrease to $\sim 1.2 \times 10^{-5}$ as conversion increases and as alcohol is formed. The sharp decrease in equilibrium ratio as the liquid phase shifted from olefin to aldehyde to alcohol neatly explained the down-stream rhodium accumulations observed in recovered catalysts.

TABLE III

RHODIUM CONCENTRATIONS IN SOLUTION
OVER AMINE RESINS
1-hexene, 100°C, 1000-2000 psig, 2:1 H_2:CO

[Rhodium]	Loading	Conversion	[Rh]/Loading
$1.3 \times 10^{-5}\underline{M}$	0.23%	40%	5.7×10^{-5}
4.0×10^{-5}	0.72%	15%	5.6×10^{-5}
5.5×10^{-5}	0.92%	30%	6.0×10^{-5}
14.0×10^{-5}	1.4%	35%	10.0×10^{-5}
12.0×10^{-5}	2.0%	20%	6.0×10^{-5}
17.0×10^{-5}	2.0%	25%	8.5×10^{-5}
			7.0×10^{-5}
0.5×10^{-5}	0.23%	65%	2.2×10^{-5}
1.5×10^{-5}	1.25%	80%	1.2×10^{-5}

Two factors were considered as possible causes for the changing equilibrium ratio, decreasing olefin content (as found with phosphine) and changing solvent (aldehyde vs. alcohol). To distinguish between the two, rhodium elution was monitored with pure aldehyde feed (1500 psig, 4:1 H_2:CO). As aldehyde hydrogenation increased, rhodium in solution declined linearly, from a [Rh]/Loading ratio of ~12 x 10^{-5} at 25% conversion to 4 x 10^{-5} at 80%. n-Heptanol itself gave a ratio of 3 x 10^{-5}. These data evidence a major solvent effect on rhodium equilibria.

As with the phosphine, pressure and temperature had an additional influence on rhodium concentrations in solution, an influence generally obscured by the effects of changing conversions (and solvent). Using pure benzene as the feed it was shown that both increased temperature and decreased CO pressure led to reduced rhodium concentrations. At 1000 psig H_2, 100°C, an increase in CO pressure from 250 to 1000 psig effected an increase in equilibrium ratio from 0.4 x 10^{-5} to 1.1 x 10^{-5}. At 2000 psig 1:1 H_2:CO, the ratio decreased from 1.1 x 10^{-5} to 0.9 x 10^{-5} when the temperature was raised from 100 to 125°C.

IV.   Discussion

Behavior of the phosphine resins can be understood in terms of the coordination chemistry reported by Wilkinson and coworkers (12). Increasing CO pressure shifts the equilibrium to favor free, eluting rhodium carbonyl, carbonyls whose stability relative to the phosphine complexes decreases with increasing temperature. That these eluting species must be of only minor catalytic importance is evidenced by the near absence of the olefin isomerization reaction over these phosphine resins.

Although these present data show a marked dependence of equilibrium rhodium concentrations on solvent and on reaction conditions, some approximate comparison is possible with a value reported for rhodium over silica-bonded phosphine complexes (10). During the hydroformylation of 1-hexene (650 psig, 1:1 H_2:CO, 140°C, high conversion), concentration in solution was 10 ppm or about 8 x 10^{-5}M.

The catalyst contained 2% rhodium, corresponding to an equilibrium ratio of 4×10^{-5} moles per liter/percent loading, but only 2 moles phosphine per mole rhodium (1.2%P). Extrapolating present data with phosphine resins to the same reaction conditions, one would predict an equilibrium ratio for the resins of about 0.9×10^{-5}, the lower value being attributed to the larger excess of phosphine ligand. Similar comparison is possible with reports for silica-bonded amines, where equilibrium ratios as low as $\sim 2 \times 10^{-5}$ are reported for donors such as triethyl amine analogs (10).

No consensus exists as yet regarding the form of the active catalyst complex in rhodium-amine hydroformylation processes. That facile hydrogenation of aldehydes to alcohols becomes possible in the presence of tertiary amines of specified pK_b and concentration has been reported by two groups (2,11), and it is known that rhodium is predominantly present as carbonyl cluster anions in the amine resin catalysts (2,8). Distribution and catalytic importance of the species, $HNR_3^+Rh_x(CO)y^-$, $R_3N-Rh_x(CO)_y(H)$, and $H-Rh_x(CO)_y$, remains however unresolved.

V.   Conclusions

Dominant in the catalytic behavior of rhodium complexes has been the influence of coordinated ligands, ligands retained in the coordination sphere throughout a catalytic cycle. The validity of such a concept and its practical utility have been probed by measuring the rates of rhodium elution from phosphine- and amine-resin hydroformylation catalysts.

For both types of resins, rhodium concentrations in solution were proportional to percent rhodium in the resin, (tested in all cases with a large excess of ligand). Rhodium elution from phosphine resins exhibited a pattern in good agreement with the behavior of the monomeric complexes, elution increasing with CO pressure or with olefin content of the effluent and decreasing with increased temperature.

Metal concentrations over amine resins, although they exhibited an equilibrium behavior, were strongly solvent dependent. Elution was particularly enhanced by high aldehyde (as compared with alcohol) in the liquid phase, an effect which obscured any potential contribution from olefin and which led to significant redistribution of rhodium within a reactor bed during use.

VI.   Acknowledgments

The encouragement and helpful suggestions of our associates and the technical assistance of G. C. Slotterback and R. E. Thomsen have been greatly appreciated.

VII.   References

1.   Haag, W. O., Whitehurst, D. D., in J. W. Hightower (Ed.), Catalysis, North-Holland, Amsterdam, 1973, p. 465.

2.   Jurewicz, A. T., Rollmann, L. D., Whitehurst, D. D., Advances Chem. Series (1974) 132, 240.

3.   Allum, K. G., Hancock, R. D., Howell, I. V., Lester, T. E., McKenzie, S., Pitkethly, R. C., Robinson, P. J., J. Organometal. Chem., (1976) 107, 393.

4.   Pittman, Jr., C. U., Smith, L. R., Hanes, R. M., J. Amer. Chem. Soc., (1975) 97, 1742.

5.   Bonds, Jr., W. D. Brubaker, Jr., C. H. Chandrasekaran, E. S., Gibbons, C., Grubbs, R. H., Kroll, L. C., ibid (1975) 97, 2128.

6.   Grubbs, R. H., Kroll, L. C., Sweet, E. M., J. Macromol. Sci. Chem., (1973) A7(5), 1047.

7.   Allum, K. G., Hancock, R. D., Howell, I. V., Pitkethly, R. C., Robinson, P. J., J. Organometal. Chem. (1975) 87, 189.

8.   Rollmann, L. D., Inorg. Chim. Acta, 6, 137 (1972).

9.   Sandell, E. B., Colorimetric Determination of Traces of Metals, Third Edition, Interscience, New York, 1959, p. 769.

10. Hancock, R. D., Howell, I. V., Pitkethly, R. C., Robinson, P. J., in Catalysis, B. Delmon and G. Jannes (Eds.), Elsevier, Amsterdam, 1975, p. 361.

11. Fell, B., and Guerts, A., Chem. Ing. Tech. Z. (1972) <u>44</u>, 708.

12. Evans, D. Osborn, J. A., Wilkinson, G., J. Chem. Soc., <u>1968A</u> , 3133.

CATALYTIC PROPERTIES OF PHOTOCHEMICALLY ATTACHED METAL CLUSTERS ON POLYVINYLPYRIDINE

Amitava Gupta and Alan Rembaum
Jet Propulsion Laboratory

Harry B. Gray
A A Noyes Laboratory of Chemical Physics

ABSTRACT

Photolysis of rhodium carbonyl ($Rh_4(CO)_{12}$) solutions in presence of suspended polyvinylpyridine coated on controlled pore glass particles produces a polymer metal complex which hydroformylates olefins under mild conditions. The rate of hydroformylation has been studied as a function of gas pressure and reactant concentrations. It has been shown in preliminary experiments that $Co_4(CO)_{12}$ solutions form a complex with polyvinylpyridine on photolysis, and this complex functions as a hydroformylation catalyst, whereas unreacted $Co_4(CO)_{12}$ does not catalyse the hydroformylation reaction under comparable conditions.

I. INTRODUCTION

Research in synthesis and characterization of metal cluster complexes has recently received impetus from demonstration that this type of complexes possesses novel catalytic properties arising partly from the presence of multiple reactive sites and partly from the steric configuration of these sites or their environment. Immobilization of these transition metal catalysts on the surface of a polymeric ligand enhances the potential of their practical application, since it makes the process of catalyst separation easier and makes the catalytic species more resistant to deactivation particularly through aggregation. For example, Grubbs[2] reported that attachment of titanocene to phosphinated polystyrene produces an olefin hydrogenation catalyst which retains its activity for longer periods. Polymer metal complexes have been studied with many goals in view, such as constructing models for enzymes[3] and improving stability of homogenous catalysts in olefin hydrogenation [4,5] and hydroformylation reactions[6,7].

A series of transition metal cluster carbonyls have been photolyzed in presence of a polymeric ligand (polyvinyl-pyridine) and the catalytic properties of the photochemically generated metal polymer complexes have been studies with respect to the hydroformylation reaction (eq 1). This reaction involves addition of hydrogen and carbon monoxide to an olefin yielding an isomeric mixture of aldehydes, which may then be reduced to the corresponding alcohols, often referred to as OXO alcohols.

$$
\begin{array}{c}
R_1 \quad\quad R_3 \\
\diagdown\quad\diagup \\
\diagup\quad\diagdown \\
R_2 \quad\quad R_4
\end{array}
+ H_2 + CO \rightarrow
\begin{array}{c}
R_1 \; HCO \; R_3 \\
\diagdown\;|\;\diagup \\
\diagup\;|\;\diagdown \\
R_2 \; H \;\; R_4
\end{array}
+
\begin{array}{c}
R_1 \; HCO \; R_3 \\
\diagdown\;|\;\diagup \\
\diagup\;|\;\diagdown \\
R_2 \; H \;\; R_4
\end{array}
\qquad \text{eq (1)}
$$

Photolysis of the metal carbonyls presumably result in either scission of metal-metal bonds or ligand dissociation, and the polymeric ligands may be expected to trap these coordinatively unsaturated species. However, in many cases there is competing thermal interaction between the metal complex and the polymer so that the following reaction modes may be distinguished:

a. Excitation of the metal carbonyl leading to its dissociation and formation of a monomeric fragment (as in the case of $Ru_3(CO)_{12}$) or ligand dissociation as has been reported to occur from $Co_2(CO)_8$.

b. Attachment of these fragments to the polymeric ligands.

c. Thermal interaction between the metal carbonyl and the polymer as is observed with $Rh_4(CO)_{12}$ and $Co_2(CO)_8$. A competition between thermal and photochemical interaction takes place with $Co_4(CO)_{12}$.

d. Photolysis of the thermally generated polymer metal complex either through direct excitation or energy transfer from metal carbonyl excited states in solution.

The rate of the reaction of tetrarhodium dodecacarbonyl with polyvinylpyridine is plotted in figure 1. The crystal structure of $Rh_4(CO)_{12}$ has been determined[8] and C-13 NMR studies indicate[9] that the bridging and terminal carbonyls are undergoing rapid interconversion at room temperature. CO is readily lost thermally at room temperature as evidenced by facile exchange with $C^{13}O^{16}$ and conversion to $Rh_6(CO)_{16}$ on being exposed to vacuum[a]. Photolysis of $Rh_4(CO)_{12}$ solutions (n-Hexane) in sealed degassed ampoules at room temperature indicates that the material is photostable. However, it is possible that the following equilibrium is established:

$$
Rh_4(CO)_{12} \; \xrightleftharpoons[\Delta]{h\nu} \; Rh_4(CO)_{11} + CO \qquad \text{eq (2)}
$$

a. Gupta, A., Unpublished Results

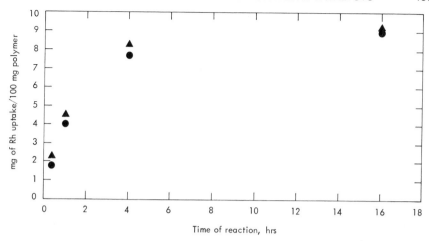

Fig. 1. *Rhodium uptake by Polyvinylpyridine as function of time;* $[Rh_4(CO)_{12}]=0.5$ *mg/ml, solvent: n-hexane, temperature: 20°C; photochemical reaction* -▲-▲-, *thermal reaction* -●-●-.

with the equilibrium lying so far to the left that no spectral change may be detected. The rate measurements (fig. 1) were carried out by measuring rhodium concentration in solution using atomic absorption spectroscopy. The rate of rhodium uptake is the same whether the reaction is thermal or photochemical. The rate of gas evolution was monitored during the reaction on a mass spectrometer, and it turns out that while there is no gas evolution when the reaction is thermal, photochemical attachment of tetrarhodium deodecacarbonyl on polyvinylpyridine generates one mole of CO per mole of Rh_4 cluster attached. The catalytic activity of the polymer-metal complex rhodium uptake by the polymer, as shown in fig. 2. The activity, normalized for rhodium content, increases dramatically as the photochemical reaction proceeds, indicating that secondary photoprocesses may occur on the polymer surface after the initial thermal interaction (which might be chemisorption) has taken place. Characterization of these rhodium polymer complexes is in progress. Resonance raman experiments will be carried out which will yield information on the structure of the metal cluster on the polymer surface. Photoaccoustic spectroscopy has been used to obtain electronic spectra of these materials in the powder form and has been described elsewhere.

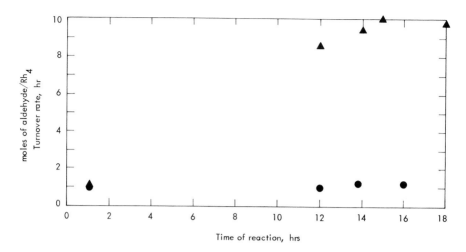

Fig. 2.   *Catalytic activity of Polymer metal complex as function of time of reaction between $Rh_4(CO)_{12}$ and polyvinylpyridine; Pressure: 90 psia, temperature: $20°C$, [alkene] = $0.5\underline{M}$, $[H_2]/[CO]$ = 80:20; photochemical reaction* -▲-▲-, *thermal reaction* -●-●-.

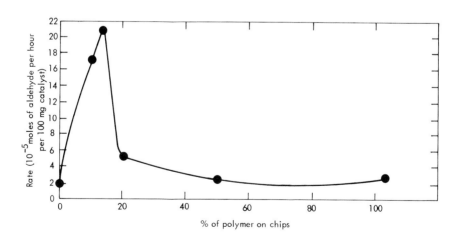

Fig. 3.   *Catalytic activity as function of polymer loading on CPG particles; conditions of hydroformylation reaction same as in Fig. 2; all catalysts were prepared photochemically, using 16 hrs. radiation period.*

II. COMPOSITION OF THE Rh-POLYMER CATALYSTS

The activity of the photochemically synthesized complex
is strongly dependant on the form of the polymer substrate.
Polyvinylpyridine has been used in three forms: 1) pure
polymer either synthesized in our laboratory or purified
commercial samples, 2) crosslinked microspheres of diameter
varying from 200 - 2000 nm, and 3) as a coating on controlled
pore glass particles, diameter varying from 37 - 500μ and pore
diameter from 1200 - 2500Å. The activity is highest when the
polymer is employed as a coating on CPG particles, presumably
because the surface area is the largest. Figure 3 shows a
plot of activity <u>vs</u> polymer loading on CPG particles. The
maximum activity corresponds to a polymer loading which would
produce a monolayer if the deposit is assumed to be uniform.
The thermally synthesized complex has no such peak in its
activity plot. The yield of aldehydes (both isomers) was
plotted as function of time using various types of polymer
substrate. The reactant concentrations were all maintained
at a level so that percent conversion was small. Figures
4 and 5 show some typical plots. These yield plots indicate
that the catalysts are stable for up to 40 hours. Other runs
have been carried out up to 80 hours without any significant
deviation from linearity.

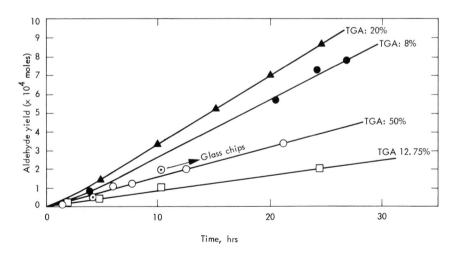

*Fig. 4. Yield of aldehydes <u>vs</u> time using polymer metal
complexes synthesized thermally; conditions same as before;
TGA: thermogravimetric analysis carried out before metal
attachment, indicates polymer loading by wt%.*

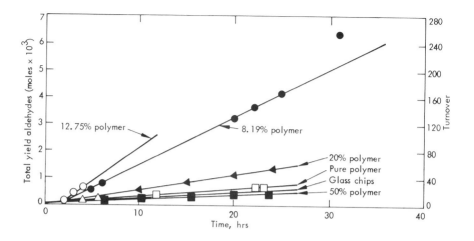

Fig. 5. Yield of aldehydes vs time using polymer metal complexes synthesized photochemically; conditions same as before.

III. PARAMETRIC STUDIES

We have measured the activity of the photochemically synthesized catalyst as a function of H_2/CO molar ratios, as shown in figure 6. The activity drops, not unexpectedly, as the mole ratio of CO increases from 1 to 10%, but then it levels off and increases slightly. Some of these runs were repeated at higher pressures. Figure 7 shows a plot of activity vs total pressure of the system. The total pressure includes the vapor pressure of the solvent (n-hexane) and the alkene (1-pentene in all cases). These and other high pressure (>1 atm) runs were carried out in a thermostatted and stirred pressure reactor. Figure 8 shows the variation of activity with alkene concentration. The run using pure alkene as the liquid phase showed significant decay in cat-alytic activity with time and the liquid phase became yellow indicating that rhodium was being leached out into the solu-tion. The isomer ratio was measured in all runs. The ratio varies with time, and the linear isomer predominates as time progresses, as shown in figure 9. This cannot be explained by assuming that the catalytic sites forming the branched isomer are decaying with time as the total yield remains linear with time over the same measurement period. It appears that the sites are undergoing a gradual change in their sterochemical environment, so that they are making more and more of the linear isomer.

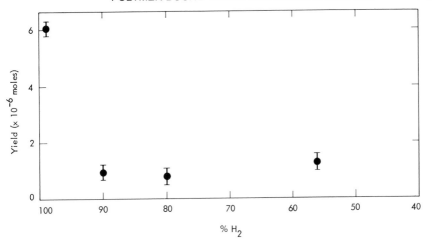

Fig. 6.   Yield of aldehydes vs H_2/CO mole ratios; pressure: 148 psia, temperature: 20°C, reaction time: 17 hours; all · catalysts synthesized by 1 hr. radiation of polyvinylpyridine and $Rh_4(CO)_{12}$ solution.

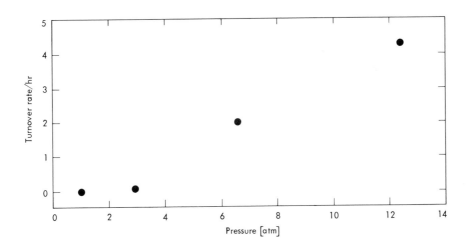

Fig. 7.   Turnover rate vs total pressure; temperature: 20°C $[H_2]/[CO]$: 80:20, [alkene]: 0.5 M; all catalysts synthesized by 16 hours initial irradiation.

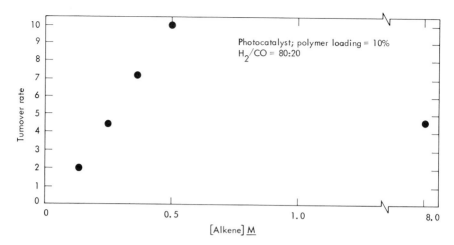

Fig. 8.   Turnover rate vs [alkene]; conditions same as before, pressure: 90 psia.

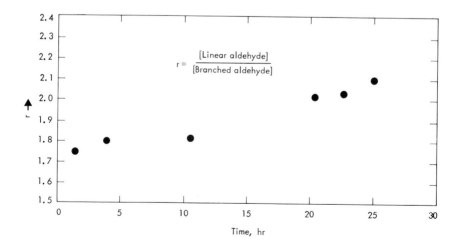

Fig. 9.   Isomer Ratio changes as catalysis proceeds; all conditions same as before, [alkene]: 0.5 M̲.

The photochemically generated complex is found to be quite specific in activity, i.e. there was no hydrogenation or isomerization of the olefins. We are studying the activity of these complexes as a function of polymer porosity and rigidity as well as surface area. The best turnover rate we have observed so far at 90 psia total pressure and 20°C with a H_2/CO mole ratio of 80:20 and 1-pentene concentration of 0.5 M is 10/hour. This is expected to go up to ca 30/hour at 175 psi and 20°C and a H_2/CO mole ratio of 50:50. These rates are comparable to those achieved by industrial rhodium catlaysts operating under more drastic conditions.

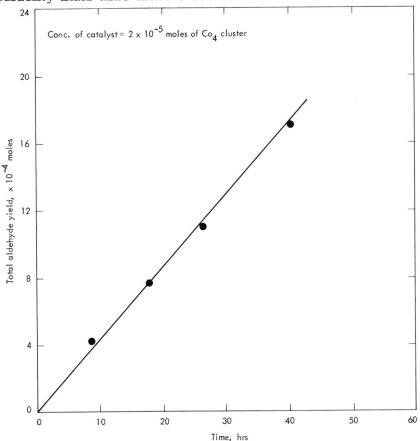

Fig. 10.   Hydroformylation using cobalt polymer complex; pressure: 150 psig, temperature: 20°C, [alkene]: 0.5 M; catalyst synthesized by irradiation of $Co_4(CO)_{12}$ solution in presence of Polyvinylpyridine (20 hours).

IV.   REACTION OF $Co_4(CO)_{12}$ WITH POLYVINYLPYRIDINE

Tetracobalt dodecacarbonyl $Co_4(CO)_{12}$ has a tetrahedral metal cluster, similar to $Rh_4(CO)_{12}$. This cluster cobalt carbonyl is considerably more stable than $Co_2(CO)_8$ and unlike $Co_2(CO)_8$ does not hydroformylate olefins stoichiometrically at room temperature in a H_2 atmosphere. The reaction of $Co_4(CO)_{12}$ solution with an olefin (1-pentene) and syngas was followed spectroscopically at an olefin concentration of 0.5 $\underline{M}$ and syngas pressure of 150 psig. It appears that two of the basal, terminal carbonyls are replaced by olefin molecules and there is no further reaction at room temperature. When a $Co_4(CO)_{12}$ solution is photolysed in presence of crosslinked polyvinyl-pyridine microspheres (diameter 5μ) a cobalt polymer complex is formed which is quite stable and can be transferred to the pressure reactor in air. Appreciable rates of hydroformylation are observed at 150 psig total pressure using a 80:20 mixture of H_2 and CO and room temperature. Figure 10 shows a typical yield plot. Parametric studies on this system are in process.

V.   REFERENCES

1.   a) Muetterities, E.L., Bull. Soc. Chim. Belg. 84, 959 (1975).
     b) Muetterities, E.L., ibid 85, 451 (1976).
2.   Grubbs, JACS.
3.   Rollman, L.D., J. Amer. Chem. Soc. 97, 2132 (1975).
4.   Bonds, W.D., Brubakar, C.H., Chandrashekharan, E.S.,
     Gibbons, C., Grubbs, R.H., and Kroll, L.C., J. Amer.
     Chem. Soc. 97, 2128 (1975).
5.   Collman, J.P., et. al., J. Amer. Chem. Soc. 94, 1789 (1972).
6.   a) Evans, G.O., Pittman, C.U., Jr., McMillan, R.,
     Beach, R.T., and Jones, R., J. Organomettalic Chem.
     67, 295 (1974).
7.   Gupta, A., Rembaum, A., Gray, H.B., and Volksen, W., 172nd
     National Meeting, American Chemical Society, San
     Francisco, 1975.
8.   Wei, C.H., Wilkes, G.R., and Dahl, L.F., J. Amer. Chem. Soc.
     89, 4792 (1967).
9.   Cotton, F.A., Kruczynski, L., Shepra, B.L., and Johnson,
     L.F., J. Amer. Chem. Soc. 94, 6191 (1972).

PHOTOACOUSTIC SPECTROSCOPY OF ORGANOMETALLIC COMPOUNDS WITH APPLICATIONS IN THE FIELDS OF QUASI-ONE-DIMENSIONAL CONDUCTORS AND CATALYSIS*

R.B. Somoano, A. Gupta, W. Volksen, and A. Rembaum

Jet Propulsion Laboratory
Pasadena, California 91103
and
R. Williams
California Institute of Technology
Pasadena, California 91125

The use of photoacoustic spectroscopy (PAS) to obtain information about the electronic absorption spectra of organometallic compounds is described. PAS is used to investigate the optical properties of: (a) several quasi-one-dimensional rhodium metal complexes and (b) several transition metal catalysts which are immobilized on polymeric microsphere substrates. The PAS spectra reveal dramatic differences between highly conducting rhodium, polymeric chains and dimeric (and nonconducting) rhodium complexes. For the catalysts, the photoacoustic spectra provide information concerning the degree of reductionof the transition metal and metal-ligand interactions which correlate with catalytic behavior.

I.   INTRODUCTION

Photoacoustic spectroscopy (PAS) is a recently developed technique (1) that measures the UV, visible, and near IR absorption spectrum of materials which are unsuitable for study by conventional optical means. The samples may be in a fine powdered form, solid, semisolid, or liquid, optically opaque, and either crystalline or amorphous. The problems associated with scattered light are of little consequence in PAS. These unique features have facilitated the study of organometallic compounds which are of importance in the fields of quasi-one dimensional (1D) conductors and photo-triggered catalysis.

*This paper represents one phase of research performed by the Jet Propulsion Laboratory, California Institute of Technology sponsored by the National Aeronautics and Space Administration, Contract NAS7-100.

Of particular interest for 1D conductors are organometallic compounds containing rhodium atoms which stack in a chain-like fashion with a uniform spacing between the metallic atoms. Such a configuration is required to obtain high electrical conductivity. However, a uniformly stacked metallic chain may lower its energy via a Peierls transition in which the chain distorts into a nonuniformly stacked configuration, e.g., dimers, trimers, etc. This distorted structure results from the creation of an energy gap at the Fermi surface, and thus, nonmetallic behavior results. In order to characterize and correlate the properties of distorted and nondistorted rhodium metal chains, we have measured the photoacoustic spectra of rhodium dimers, bridged dimers, and polymeric chains.

In the field of catalysis, it is known that the reaction of transition metal species which are homogeneous catalysts, with polymeric ligands may lead to the formation of anchored catalysts of greater stability and activity. We have reported the formation of a rhodium (Rh) - polyvinylpyridine (PVP) complex through photolysis of $Rh_4(CO)_{12}$ solution in the presence of PVP(2). This complex is an active olefin hydroformylation catalyst. It is necessary to determine the electronic spectra of these metal-polymer complexes in order to elucidate the structure-reactivity relationships. We describe results of PAS on several of these anchored catalysts.

II.   THE PHOTOACOUSTIC (PA) EFFECT

For PAS, a sample is enclosed in an air tight, nonresonant cell containing air at atmospheric pressure and a sensitive microphone. The sample is illuminated through a quartz window with periodically modulated monochromatic radiation. The PAS signal originates when electrons, excited during absorption, decay nonradiatively, thereby creating a periodic heat distribution at the point of absorption (Fig. 1). This heat distribution diffuses to the sample-air interface where it gives up its thermal energy to the adjacent thin boundary layer of air. The periodic expansion and contraction of this boundary layer acts as an acoustic piston on the remaining air space, and thus, creates pressure

PHOTOACOUSTIC EFFECT

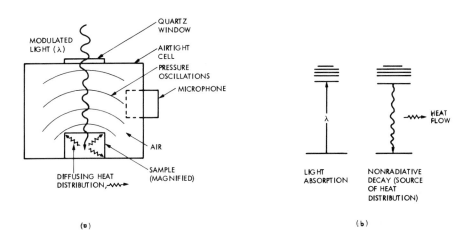

Fig. 1. Photoacoustic process and sample cell configuration

oscillations which are detected by the microphone.
Samples of mass ≥ 1 mg and with optical absorption
coefficients of $\overline{0} - 10^4$ cm^{-1} may be studied. The
samples used in the present study were in the form
of fine powders.

III. RESULTS

Organometallic semiconductors are of interest
in the field of quasi-one-dimensional (1D) conductors. The compounds consist of weakly coupled
chains of molecular units and may exhibit very high
electrical conductivity parallel to the chains.
The compounds may be metallic if the atoms or molecules are uniformly spaced along the chain. However, quasi-one-dimensional metallic chains are
fundamentally unstable with respect to certain lattice distortions (Peierls' Instability). Thus, in
some compounds the chains will distort in such a
manner that the atoms or molecules form dimers,
trimers, tetramers, etc. This distortion, or transition, can occur at temperatures above or below
room temperature. The compounds with distorted
chains are semiconductors since the nonuniform
spacing of the molecules along the chain opens up a
gap at the Fermi level in the electronic energy

spectrum (Fig. 2). Thus, the electrical and magnetic properties are crucially dependent upon the
chain structure. The compounds are usually darkly
colored powders. Single crystals, which are difficult to obtain, are in the form of very small
needles (a few mm long) making conventional optical
studies very difficult.

QUASI-ONE-DIMENSIONAL CONDUCTORS

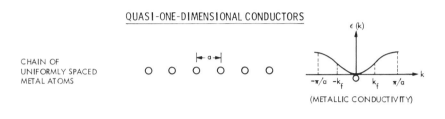

PEIERLS' INSTABILITY

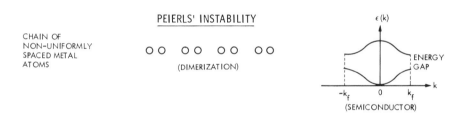

Fig. 2. Schematic of the Peierls' Instability in quasi-
one-dimensional conductors.

PAS may be utilized to gain insight into the
structural characteristics of these materials.
Fig. 3 shows the PA spectra of several rhodium
organometallic compounds. Fig. 3(a) shows the PA
spectrum of an interesting rhodium "bridged" dimer
compound (3), in which the rhodium atoms are physically bound into a dimeric structure by bridging
ligands. The bridged dimers are not part of a
chain structure in this compound. However, the
interesting feature is the presence of the strong
absorption band at ∿575 nm which is due to the
dimeric structure (3). A similar band is found in
the solution spectrum. Figures 3(b) and 3(c) show
PA spectra of two rhodium chain compounds. The
rhodium atoms are square planar coordinated by iso-
cyanide ligands (R-CN), and stack to form rhodium
chains in contrast to the rhodium "bridged" dimer.

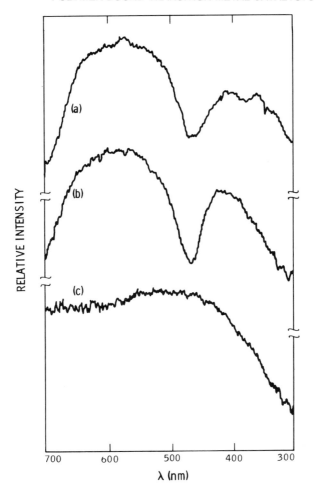

Fig. 3. Photoacoustic spectra of several rhodium organometallic compounds. (a) rhodium "bridged" dimer, (b) rhodium (phenylisocyanide)₄ tetraphenylborate, [Rh(φ-CN)₄] Bφ₄. The absorption band at ∿575 nm in Fig. 3(a) and 3(b) is associated with the formation of dimers. (c) rhodium (vinylisocyanate)₄perchlorate, [Rh(VI-CN)₄]ClO₄.

Nevertheless, the PA spectra of rhodium (phenyliso-cyanate)₄tetraphenylborate, Fig. 3(b), reveals the same strong rhodium-rhodium dimer band at ∿575 nm indicating a dimeric chain structure. The room temperature electrical conductivity of this material is quite low ($\sigma_{RT} < 10^{-10}\Omega^{-1} - cm^{-1}$) as would be

expected for the nonuniform (dimeric) spacing along
the chain. In contrast, the PA spectrum of rhodium
(vinylisocyanide)$_4$perchlorate, Fig. 3(c), does not
reveal any sign of the dimer, suggesting that this
material contains chains of uniformly spaced rhodium
atoms. Indeed, the electrical conductivity of this
compound ($\sigma_{RT} \approx 2\Omega^{-1} - cm^{-1}$) is the highest known for
any rhodium chain complex (4). All of the struc-
tural information deduced from the PA spectra of
Fig. 3 has been fully confirmed by a single crystal
X-ray diffraction studies (4). Similar results have
been obtained on iridium chain complexes (5).

In summary, PAS provides a very simple technique
for gaining insight into the structural and electrical properties of quasi-1D conductors without
having to grow single crystals or perform X-ray
measurements. In this way, many new potentially
conducting compounds may be screened without the
expenditure of excessive time and funds.

An important area where PAS is being utilized at
present is in the study of photochemical processes
and catalysis. An example involves the investigation of the reaction of transition metal complexes
with polymeric ligands to form anchored catalysts
which have been used to catalyze hydrogenation and
hydroformylation of olefins. PAS has been used to
investigate the electronic structure of these
metal-polymer complexes in order to elucidate chemical processes and structure-reactivity relationships. An example of a PA study of a model catalyst
system involving a well characterized compound,
tungsten hexacarbonyl $W(CO)_6$, is shown in Fig. 4.
The PA spectrum of $W(CO)_6$ is shown both before and
after photochemical reaction with polyvinylpyridine.
Fig. 4(a) shows the PA spectrum of $W(CO)_6$ and
reveals the singlet-triplet ligand field transition
at $\sim$350 nm as well as the corresponding singlet-
singlet transition at 310 nm (6). Fig. 4(b) shows
the PA spectrum of $W(CO)_5L$, where the ligand L is
NH_3. This compound is formed upon irradiation of
$W(CO)_6$ in the presence of the ligand and is shown
to reveal the effect of ligand substitution in
$W(CO)_6$. The characteristic singlet-triplet and
singlet-singlet transitions at 457 and 416 nm,
respectively, of $W(CO)_5NH_3$ (6) are observed in the
form of a broad absorption band from 375 to 475 nm.

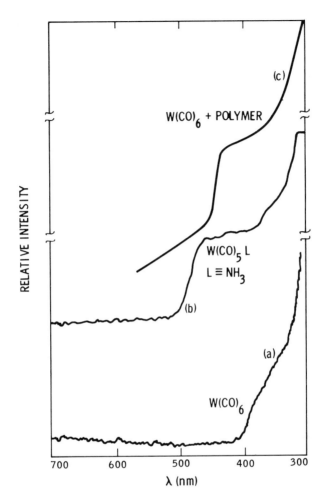

Fig. 4.   *Photoacoustic spectra of tungsten-hexacarbonyl,*
W(CO)$_6$, complexes.   (a) W(CO)$_6$, (b) W(CO)$_5$NH$_3$, (c) W(CO)$_6$ +
polyvinylpyridine.   The complexes in (b) and (c) are
photoproducts.

Fig. 4(c) shows the PA spectrum of the photoproduct
of W(CO)$_6$ and polyvinylpyridine and clearly reveals
the W(CO)$_5$-pyridine absorption band similar to that
seen in Fig. 4(b).   The polymer anchored W(CO)$_5$
species has been observed to catalyze olefin isomerization and hydrogenation.   Thus, PAS may be
used to study and characterize solid state photoproducts where conventional optical and structural
(i.e., X-ray) techniques would be totally inadequate.

An example of a less well understood, but potentially useful polymer anchored catalyst involves the
transition metal cluster complex, tetrarhodium dodecacarbonyl, $Rh_4(CO)_{12}$. Fig. 5 shows the PAS of:
(a) $Rh_4(CO)_{12}$, (b) the photoproduct of the reaction
of $Rh_4(CO)_{12}$ solution with polyvinylpyridine, and
(c) the thermal product (i.e., the same as (b) but
not exposed to UV radiation). Both of the polymer-

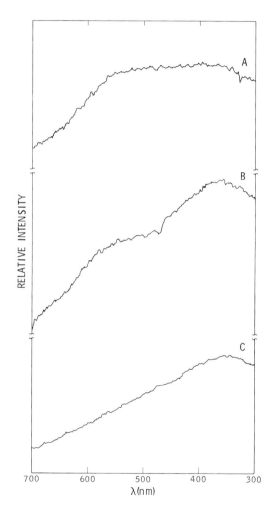

*Fig. 5. Photoacoustic spectra of tetrarhodium dodecacar-
bonyl, $Rh_4(CO)_{12}$,-polymer complexes. (a) $Rh_4(CO)_{12}$, (b) the
photoproduct of the reaction of $Rh_4(CO)_{12}$ solution with polyvinylpyridine, and (c) same as (b) but not irradiated with UV
light (i.e., a thermal product).*

metal complexes, (b) and (c), are found to be hydroformylation catalysts, but the phototriggered catalyst, (b), is considerably more active. This extra
activity is associated with the absorption band
(shoulder) at 600 nm in Fig. 5(b) since the less
active thermal catalyst lacks this absorption.
Upon prolonged exposure to air, the phototriggered
catalyst decomposes as indicated by the loss of the
600 nm absorption. Its PA spectrum then resembles
that of the thermal catalyst and exhibits correspondingly less catalytic activity. Similar studies are
being carried out with $Co_4(CO)_{12}$.

Another use of PAS in this field has been to
detect the degree of reduction of metals reacted on
polymeric substrates. By using PAS to monitor the
reaction of, say, chloroplatinate, K_2PtCl_4, with
polyvinylpyridine, one may easily tell when the
chloroplatinate has been fully reduced to platinum
metal. The use of PAS in catalysis studies is just
starting, but there is every indication that it will
prove to be a valuable tool.

In summary, PAS offers a novel way to obtain
information on the electronic spectra of many materials of interest in the fields of 1D conductors and
catalysis. PAS appears quite likely to become a
valuable analytical tool and new uses for the technique in various scientific fields are emerging at
a rapid rate.

IV. REFERENCE

1. Rosencwaig, A., Analy. Chem., **47**, 592A (1975)
    and references therein.

2. Gupta, A., Rembaum, A., Slusser, P., Volksen,
    W., and Gray, H.B., 172nd ACS National Meeting,
    San Francisco, Ca. (1976).

3. Lewis, N.S., Mann, K.R., Gordon, J.G., and
    Gray, H.B., J. Am. Chem. Soc., **98**, 7461 (1976).

4. Williams, R., Hsu, C.H., Cuellar, E., Gordon,
    J.G., Samson, S., Hadek, V., and Somoano, R.,
    172nd ACS National Meeting, San Francisco, Ca.
    (1976).

5. Rosencwaig, A. Ginsberg, A.P., and Koepke, J.W., Inorg. Chem. $\underline{15}$, 2540 (1976).

6. Wrighton, M.S., Morse, D.L., Gray, H.B., and Ottesen, D.K., J. Am. Chem. Soc., $\underline{98}$, 1111 (1976).

ANTIFOULING APPLICATIONS OF VARIOUS TIN-
CONTAINING ORGANOMETALLIC POLYMERS

by

W. L. Yeager and V. J. Castelli

David W. Taylor Naval Ship Research and Development Center

ABSTRACT

Various tin-containing Organometallic Polymers
(OMPs) have been developed to provide more ecologically
compatible alternative marine antifoulants than currently available tin-based antifouling coatings. These OMPs
are unique because they employ a marine pesticide
chemically bound to a polymer backbone resulting in the
controlled release of the pesticide into the environment.
Preliminary results show that these polymers have considerably less environmental impact than their related
state-of-the-art marine pesticides while still maintaining fouling-free performance after four years of static
panel immersion tests. Currently, four polymers exhibiting good antifouling performance are undergoing
formulations into antifouling paint systems. It has
been found that formulation problems peculiar to each
of the four polymers could possibly be related to the
physical properties of the polymers themselves. Preliminary polymer characterization has given insights
into these problems which eventually will lead to
further polymer reformulation and property improvement.

I. INTRODUCTION

The problems of conventional antifouling coatings that
leach excessive amounts of toxicant into the environment are
well known. In addition to their pollution impact, they
possess relatively short service lives, with 18 - 36 months
maximum. The desire to solve these problems prompted the
Navy to search for a new and more efficient marine antifouling system. One system presently under investigation
involves the use of various tin containing Organometallic
Polymers (OMPs). OMPs employ a polymer backbone to which
marine pesticide groups are chemically attached. Chemically bonding the marine pesticide to a polymer backbone
results in controlled toxicant release from the film. Controlled zeroth order release as an attainable polymer design
goal, effectively eliminates the problems of excessive
pesticide release, while providing desired characteristics

such as extended antifouling performance.

II.   POLYMER DESIGN THEORY

As mentioned before, the driving force leading to the
development of OMPs was the desire to effectively limit the
amounts of toxic materials released into the marine environment that occurs in state-of-the-art antifouling systems.
This can be achieved by chemically bonding a marine pesticide
to various polymer backbones. At least for the linear backbone polymers, the pendant marine pesticide is theorized to
be released into the marine environment via zeroth order
hydrolysis rather than dissolution from an insoluble matrix
as in conventional antifouling systems. The remaining
polymer matrix, minus the marine pesticide, would then
possess sufficient water solubility to dissolve , exposing
fresh, toxic polymer for further hydrolysis. This regenera-
tive process results in slow continuous release of pesticide
approaching observable zero order kinetics which provides long
term antifouling performance with minimal environmental impact.
Available experimental data for some polymer compositions
seems to indicate that this may indeed be the case. Preliminary leaching rate studies on some of the initial linear
acrylic polymers have shown a significant reduction in
toxicant release (1) at near constant rate compared to conventional antifouling coatings.

Confirming bioassay experiments utilizing these polymers
to study their effects on specific marine organisms have revealed variations in toxin release rate between different
polymer compositions. Results show dramatically longer test
animal survival times for the OMPs than with state-of-the-art
organotin paint (1a). Such information is crucial to second
generation OMP development, where upgrading of the physical
properties of the polymers while retaining good antifouling
performance will be the primary concern. In addition,
attempts will be made to achieve a balance between polymer
physical properties and tailoring of molecular weight
distributions to insure the dissolution of toxicant depleted
polymer backbone, but not unhydrolyzed polymer.

III.   POLYMER CHARACTERIZATION VS. PAINT FORMULATION

At the present time, four of the most promising OMPs
which have remained fouling free after four years of static
panel-immersion antifouling tests at Pearl Harbor, Hawaii are
now being formulated into antifouling paint systems. Formulation represents the final developmental step which converts
these polymers from laborabory curiosities to practical
products. The actual process of formulating these polymers
into paint systems has generated a unique set of problems

directly related to the chemical and physical properties of
these polymers. The four polymers that will be referred to
throughout this paper are listed below:

Polymer	Composition
M 1	Poly (Tri-n-butyltin Methacrylate-co-Tri-n-propyltin Methacrylate-co-Methyl Methacrylate) 1:1:1 mole ratio
P 43	Poly (Tri-n-butyltin Methacrylate-co-Methyl Methacrylate) 1:1 mole ratio
P 37	Tri-n-butyltin ester of Poly (Methyl Vinyl Ether-co-Maleic Acid)
P 13	Tri-n-butyltin ester of crosslinked Poly Methacrylic Acid

Preliminary characterization has shown that polymers P 37 and
M 1, and P 43 have molecular weights as listed in the
following table:

Polymer	M_n	M_w	M_w/M_n
M 1	134,000	131,600	0.98
P 37	1,430,000	---	---
P 43	109,000	160,800	1.46

As can be seen from this data, M 1 and P 43 have fairly small
polydispersities with molecular weights over 100,000. The
polydispersities of these polymers differ considerably from
an essentially monodisperse M 1 to a much wider relative
molecular weight distribution in P 43. M 1 and P 43 are
similar not only in chemical composition but in comparable
physical properties. This would indicate that small molecular weight changes have little effect on the properties of
the polymers, allowing less emphasis on quality control, with
associated lower products costs, in eventual large scale
polymer manufacture. On-going Gel Permeation Chromatography
analysis of the polymers should confirm the narrow molecular
weight distribution indicated from the determinations of M_n and
M_w. These chracteristics could be responsible for the fair
film forming properties and reasonable solubilities exhibited
by the polymers.

On the other hand, P 37 exhibits very poor film forming
properties, is marginal soluble in ketones only, and retains
surface tack at room temperature. This is probably due to
the starting polymer from which P 37 was synthesized which

was a commercially available polymer used as a detergent
stabilizer. The addition of tri-n-butyltin imparted outstanding antifouling characteristics to the polymer but did
very little to improve physical properties. Studies have
shown that molecular weights for P 37 are well over 1,000,000.
The resulting large polydispersity would contribute greatly
to solubility problems and tackiness encountered with this
polymer.

Finally, P 13 is so heavily crosslinked that it is
insoluble in all known solvents and each bead is essentially
composed of one gigantic molecule. It was synthesized from
a commercial ion exchange resin and it is being investigated
for use as a pigment for paint systems and seems to be well
suited for this purpose. Compatibility problems with resin
systems commonly used in the marine environment are absent.

The above mentioned physical properties of the polymers
resulted in different paint formulation schemes for each
polymer. During initial formulation stages, M 1 and P 43 were
compatible with rubber resins but compatability was dependent
on the relative volatility of the solvent system used. In the
case where the major solvent was the less volatile, the
smaller percentage of the more volatile solvent evaporated
more quickly, raising the amount of less volatile solvent
left in the film. This resulted in the stronger solvent holding the components in solutions and producing a uniform dry
film. On the other hand, when the more volatile solvent was
in excess, the less volatile solvent was always in lower concentration in the film even though it evaporated slower.
Continued evaporation caused depletion of the less powerful
solvent in the remaining wet film, resulting in crystallization of base paint resin, phase separation, and an undesirable
paint film. After solution of the compatibility problems, formulation moved into the more specific areas of film pigmenta-
tion and dry film physical property improvement was attempted
with standard compounds such as carbon black and titanium
dioxide. These pigments were easily dispersed in the OMP
rubber systems and produced a film with excellent leveling
characteristics and good hiding power, but primer adhesion and
dry film hardness were still marginal. After considerable
experimentation, it was found that one of the best methods of
improving hardness involved cross-linking the rubber base. A
number of agents were tried and a candidate was found that
greatly increased hardness while imparting additional water
resistance to the film. Future work will concentrate on
improving the characteristics of the OMP-rubber films utilizing different cross-linking approaches, pigments and
resin variations.
Unfortunately the formulation scheme for P 37 is much
less encouraging. Although P 37 was found to be soluble in a

number of solvents (cyclohexanone, decane, xylene, n-butyl alcohol and butylacetate), it was found to be only marginally compatible with certain rubber resins. Films generated from these systems possessed excessive drying times, poor adhesion and dry film hardness, excessive cobwebbing during spray application and retention of surface tack when dry. Attempts to remedy these undesirable properties have been largely unsuccessful resulting in film improvement only a very low OMP to binder ratios. Since reduction of OMP levels as a means to improve film properties would almost certainly result in a reduction of the antifouling performance of the resultant paint, consideration is being given to discontinuing work on P 37 in the hopes of more promising results from the three remaining polymers.

The fourth polymer, P 13, is proving to be the most versatile of the OMPs undergoing paint formulation. This material is produced in the form of off-white, insoluble granules approximately 0.25 to 0.50 mm in diameter. Because P 13 is essentially insoluble, efforts were concentrated in the areas of particle size reduction with incorporation of this material as a pigment in conventional paint systems. This allows higher toxicant levels for the various formulations while effectively eliminating resin-polymer compatibility problems. The ground polymer was substituted for conventional pigments such as TiO_2 in various paint formulations and satisfactory film properties were obtained from formulations using conventional resins and those already containing other OMPs.

From the whole paint formulation scheme, various trends can be seen. First and most important, there is a possibility for further improvement of the physical properties of the polymers to chemically tailor them for hull coatings. For instance, M 1 and P 43 exhibit fair film forming characteristics but produce hard paint films only with the addition of special hardening agents. It has been suggested that a slight degree of cross-linking or copolymerization of the acrylic resin with any of a series of vinyl monomers, would impart added strength to the film. Modification of P 37 will undoubtedly use the same rationale to provide the polymers with greater compatibility with present paint formulation schemes. This poly (methyl vinyl ether-co-maleic acid) polymer was selected for paint formulation strictly on the basis of its antifouling performance and not its structural characteristics. It appears that P 37 possesses a rather wide molecular weight distribution which could be responsible for its low solvent compatibility. As mentioned before, polymer P 13 shows the best paint formulation properties to date and will remain essentially unchanged except for particle size reduction. It will probably find wide use as a filler-pigment material.

IV. TESTING AND EVALUATION

Shortly, antifouling paints formulated from these polymers are scheduled to begin strip testing on Navy ships.
Some of these paints now have four months of fouling-free
performance in static panel testing at Miami, Florida. If
this trend continues and the goal of a five year antifouling
coating is attained, this will result in an estimated annual
cost saving of approximately $25 million which is the price
the Navy now pays for wasted fuel due to fouling.

V. REFERENCES

1. Dyckman, E.J., and Montemarano, J. A., "Antifouling
    Organometallic Polymers: Environmentally Compatible
    Materials, David W. Taylor Naval Ship R&D Center Report
    #4186, Feb 1974.

la. Private Communication, D.K. Christian, David W. Taylor
    Naval Ship R&D Center, Feb 1977.

VI. BIBLIOGRAPHY

1 - Castelli, V. J. and Yeager, W. L. "Organometallic Polymers;
    Development of Controlled Release Antifoulants."
    Proceedings of the American Chemical Society Symposium on
    Controlled Release Polymeric Formulations. (April 1976)

2. Montemarano, J. A., Cohen, S. A., and Fischer, E. C.,
    "Fouling, The Problem and Its Control by Polymeric
    Formuatlions" Proceedings of the International Controlled
    Release Pesticides Symposiums: University of Akron,
    Ohio (September 1975)

3 - Montemarano, J. A., and Dyckman, E. J., "Performance of
    Organometallic Polymers as Antifouling Materials."
    Proceedings of Los Angeles Meeting of the American
    Chemical Society, Organic Coatings and Plastic Chemistry
    Division, April 1974.

4 - Montemarano, J. A., and Dyckman, E. J., "Biologically
    Active Polymeric Materials Exhibiting Controlled Release
    Mechanisms For Fouling Prevention." Proceedings of the
    Controlled Release Pesticide Symposium. (September 1974).

5 - Yeager, W. L., and Castelli, V. J., "Development and
    Testing of Organometallic Polymer (OMP) Antifouling Paint
    Systems. Proceedinggs of American Society of Naval
    Engineers ASNE Day 1977, May 1977. (Not yet published)

COPOLYMERIZATION OF TRI-n-BUTYLTIN ACRYLATE AND
METHACRYLATE WITH FUNCTIONAL GROUP MONOMERS--
CROSSLINKING AND BIOTOXICITY OF COPOLYMERS

R. V. Subramanian, B. K. Garg and Jaime Corredor
Department of Materials Science and Engineering
Washington State University

*An approach to obtaining thermoset organotin poly-
mers, which permits control of crosslinking sites and
of polymer properties, is reported. Tri-n-butyltin
acrylate and methacrylate were prepared and copolymerized with vinyl monomers containing an epoxy or a
hydroxyl group. The experimental values of reactivity ratios were used to indicate the distribution of
the units of a particular monomer in the polymer chain.
Thus, a suitable copolymer containing either the blocks
of organotin monomer units, blocks of functional-group-
containing vinyl monomer units, or randomly distributed
units of either monomer could be selected as desired.
Varying crosslink density and rigidity were achieved
with a number of aliphatic and aromatic amines and also
with uranyl nitrate catalyst. Interestingly, evidence
for the autocatalytic effect of organotin groups was
observed in the self-curing of copolymers. Biotoxicity
data were obtained by studying the inhibition of marine
and soil bacteria and a soil fungus. The nature and
degree of crosslinking have significant effect on the
size of inhibition zones. The significance of amine
and hydroxyl groups present to the rate of hydrolysis
and leaching of organotin ester is indicated.*

I.   INTRODUCTION

Recent studies in our laboratories have established the
usefulness of crosslinking reactions in improving physical
properties of polymers containing toxic trialkyltin groups
chemically bound to them (1,2). Thermoset, epoxy polymer
based organotin network structures obtained in these studies
retained a very satisfactory antifouling action while substantially improving film properties of the polymer. The

synthetic scheme adopted in these studies was the partial
esterification by bis(tri-n-butyltin) oxide of polymers
carrying a carboxyl group, such as the 1:1 copolymer of
methyl vinyl ether with maleic anhydride, followed by a
crosslinking reaction of the free acid or anhydride groups
with epoxide monomers.

We now report a different approach to obtaining thermoset polymers containing trialkyltin groups, which permits
control of the distribution of crosslinking sites and hence
of the physical properties of the resulting structures. The
basic scheme is to prepare tri-n-butyltin esters of acrylic
and methacrylic acids, copolymerize these with vinyl monomers
containing an epoxy or a hydroxyl group, and calculate the
reactivity ratios for the monomer pairs. Since the reactivity ratios indicate the distribution of the units of a particular monomer in the polymer chain, a suitable copolymer
containing either blocks of organotin monomer units, blocks
of functional-group-containing vinyl monomer units, or randomly distributed units of either monomer can be selected as
desired. The copolymer is next crosslinked with suitable
crosslinking agents such as amines.

The effects of crosslinking and the nature of crosslinks
upon the biocidal properties of the crosslinked polymers was
also studied.

II.   EXPERIMENTAL

A.   Materials

Monomers, glycidyl acrylate and glycidyl methacrylate,
were obtained from the Borden Chemical Co., Philadelphia,
Pennsylvania, and were purified by vacuum distillation before
use. N-methylolacrylamide was supplied by the American
Cyanamid Co., Wayne, New Jersey, and was purified by recrystallization before use. All other necessary chemicals were
obtained from their respective commercial suppliers and used
without further purification.

B.   Preparation of Organotin Monomers

The tri-n-butyltin acrylate and methacrylate monomers
were prepared according to the method of Montermoso et al.
(3) through esterification of the corresponding acids. The
water of reaction was removed, through azeotropic distillation, and benzene removed from the reaction mixture by evaporation. In the case of tri-n-butyltin methacrylate, the
temperature was maintained below 30°C. The resulting pale

yellow liquid was diluted with petroleum ether, cooled to
-20°C and maintained at this temperature for 2 hr. The
product separated in stoichiometric yield as long, thick,
needle-like crystals having a melting point of 18.1-18.3°C.
The tin content of tri-n-butyltin methacrylate monomer was
determined by oxidation to tin oxide using sulfuric acid (4),
and found to be 31.4% against a calculated value of 31.7%.
Tri-n-butyltin acrylate was prepared similarly except that it
was not necessary to maintain a low temperature during preparation and purification. The needle-like tri-n-butyltin
acrylate crystals obtained in nearly stoichiometric yield had
a melting point of 74.0-74.5°C and a tin content of 32.7%,
against a calculated value of 32.9%.

C.   Copolymerization

     The copolymers were obtained by solution polymerization
using a free-radical initiator. Since the differential form
of the copolymer equation was to be used for calculating
reactivity ratios, it was necessary to limit conversion to
less than 10% in every case. The composition of the copolymer
obtained was calculated by determining its tin content. The
ratios of monomers in the initial reaction mixture were
varied over as wide a range as possible. The specific copolymerization procedures are described below.

D.   Copolymerization of Tri-n-butyltin Acrylate (TBTE-AA)
     with Glycidyl Methacrylate (GMA); and with Glycidyl
     Acrylate (GA)

     Predetermined amounts of TBTE-AA and GMA (or GA) were
placed in a small glass bottle and benzene, as solvent, was
added to obtain a concentration of monomers equal to 1.57
mole/liter in solution. The solution was next flushed with
nitrogen for 10 min, and benzoyl peroxide (1 mole % based on
total monomer moles in solution) was added as a free-radical
initiator. The bottle was capped with a serum stopper, and
placed in a constant-temperature oil bath maintained at
75 ±0.1°C. The copolymer formed was precipitated out by
pouring the reaction solution into methanol, washed with fresh
methanol, dried, and weighed.

E.   Copolymerization of Tri-n-butyltin Methacrylate (TBTE-MA)
     with Glycidyl Methacrylate (GMA); and with Glycidyl
     Acrylate (GA)

Since TBTE-MA is liquid at room temperature, it was not necessary to add benzene to obtain a homogeneous solution. The copolymerization procedure was exactly the same as described above, except that methanol-water (80:20 by volume) mixture was used to isolate the copolymer from the reaction solution.

F.   Copolymerization of N-methylolacrylamide (MAM) with Tri-n-butyltin Methacrylate (TBTE-MA); and with Tri-n-butyltin Acrylate (TBTE-AA)

The temperature of copolymerization was 65°C, and methanol was used as the reaction medium. Otherwise, the procedure for copolymerization was similar to that described above. The method of isolating the copolymer was to pour the reaction mixture into water. The precipitate, which was free of MAM monomer, was separated, dried, and redissolved in methanol. Benzene was next added to this solution to isolate the copolymer as a precipitate.

G.   Crosslinking

Determination of the copolymer composition vs. monomer ratio for the copolymerization of TBTE-MA with GMA showed that azeotropic copolymerization occurred at 46:54 molar ratio of TBTE-MA to GMA in the initial reaction mixture. The advantage of azeotropic copolymerization is that the instantaneous composition of the copolymer remains the same regardless of conversion. Thus, the 46:54 copolymer of TBTE-MA with GMA (azeotropic copolymer) was obtained in sufficient quantity by conducting the polymerization to high conversions. This copolymer was dissolved in benzene for use in crosslinking reactions.

H.   Curing with Amines

Samples (5.25 g) of azeotropic copolymer dissolved in benzene were mixed with 75, 85, 100, and 110% of the required stoichiometric amounts of diethylenetriamine (DETA) and triethylenetetramine (TETA) in aluminum dishes. Since the aliphatic amines react rapidly with epoxy resins at room temperature, all samples were cooled to 0°C before addition of amine. The benzene was allowed to evaporate at room temperature for 6 hr. Finally, the samples were cured overnight in an oven at 60°C. The aromatic amines used were m-phenylenediamine (MPDA) and 4,4´-methylenedianiline (MDA). In these

systems benzene was removed by placing the samples in an oven
at 50°C overnight and curing conducted at 150°C for 5 hr.

I.   Curing with Uranyl Nitrate

    One 5.25-g sample of azeotropic copolymer dissolved in
benzene was cured using 0.2% by weight of uranyl nitrate as
catalyst. The uranyl nitrate was dissolved in acetone at room
temperature, mixed with copolymer-benzene solution, which was
then left in an oven at 50°C overnight. The temperature was
next raised to 100°C for 6 hr and, finally, to 120°C for
24 hr.

J.   Self-Curing of the Copolymer

    The copolymer was tested for curing without the aid of
added catalyst or curing agent by subjecting a 2.5-g sample of
it to 50°C overnight, 100°C for an additional 6 hr, and
finally, 120 C for 24 hr.

K.   Characterization of Crosslinked Copolymers

    All crosslinked copolymers obtained were inspected for
tackiness, hardness, and toughness. They were next solvent-
extracted with refluxing benzene for 100 hr each in a Soxhlet
extractor to remove the soluble portion. The insoluble portion was dried, weighed to determine soluble fraction in the
crosslinked product, and finally, the tin content of the insoluble portion was determined.

L.   Biotoxicity of the Crosslinked Copolymer

    The inhibition of three bacterial cultures by azeotropic
copolymer and insoluble portions of some of the crosslinked
copolymers obtained above was studied using the culture plate
method. The three media and test organisms were 1) Nutrient
agar and BBL; Sarcina lutea, a soil bacterium; 2) Potato dex-
trose agar and 1% yeast extract; Glomerella cingulata, a soil
fungus; and 3) Nutrient agar and 3% NaCl; Pseudomonas nigrifaciens, a marine bacterium. The medium was prepared, sterilized, and allowed to cool to 37-38°C. The test organism was
then added to the cooled medium in concentrations of approximately 1 ml of 18 hr shake-flask culture of organism for every
10 ml of medium. The medium was then poured into sterilized
petri plates in 10-ml aliquots. The plates were stored in the
cold-room at 4°C for solidification to inhibit bacterial

growth. Next, weighed crosslinked copolymer samples were
placed in the center of prepared petri plates which were then
returned to the cold-room. The plates were next incubated in
the cold-room for specified times to permit diffusion of toxin
through the medium. After incubation the samples were allowed
to develop at room temperature until the emergence of test
sample inhibition zones. The plates were read for size of
inhibition zones after they had developed completely. The
development times varied with the test organism.

III. RESULTS AND DISCUSSION

A.   Copolymerization and Copolymerization Reactivity Ratios

The two organotin vinyl monomers viz. tri-n-butyltin
acrylate and tri-n-butyltin methacrylate, were each copolymerized with two epoxy-group-containing monomers, namely,
glycidyl acrylate and glycidyl methacrylate. They were also
copolymerized with one hydroxyl-group-containing monomer, N-
methylolacrylamide.   The details of the determination of
reactivity ratios for these pairs of monomers as well as the
Q,e values for each of the organotin monomers have been
discussed elsewhere (5). Briefly, the reactivity ratios
obtained for the copolymerization of tri-n-butyltin acrylate
(M_1) with glycidyl acrylate (M_2) were $r_1 = 0.295 \pm 0.053$, $r_2 =$
1.409 ± 0.103; with glycidyl methacrylate (M_2) they were $r_1 =$
0.344 ± 0.201, $r_2 = 4.290 \pm 0.273$; and with N-methylolacrylamide
(M_2) they were $r_1 = 0.977 \pm 0.087$, $r_2 = 1.258 \pm 0.038$. Similarly, for the copolymerization of tri-n-butyltin methacrylate
(M_1) with glycidyl acrylate (M_2) they were $r_1 = 1.356 \pm 0.157$,
$r_2 = 0.367 \pm 0.086$; with glycidyl methacrylate (M_2) they were
$r_1 = 0.754 \pm 0.128$; $r_2 = 0.794 \pm 0.135$; and with N-methylolacrylamide (M_2) they were $r_1 = 4.230 \pm 0.658$, $r_2 = 0.381 \pm 0.074$.
The copolymerization of tri-n-butyltin methacrylate with
glycidyl methacrylate was found to show azeotropic copolymerization at a 46:54 mole ratio. It was found that, for this
particular case, the experimental results deviated significantly from the terminal model of binary copolymerization.
The deviation tended to suggest that there was a significant
charge transfer copolymerization occurring in this case. This
aspect of the participation of organotin vinyl monomers in
charge transfer copolymerizations is being further investigated in our laboratory.
From the values of reactivity ratios it is seen that tri-
n-butyltin methacrylate monomer is much more reactive than the
tri-n-butyltin acrylate monomer. Since the reactivity ratios
vary widely, it is possible to obtain three kinds of important
copolymers. These are (a) copolymers containing blocks of

organotin units separated by random distribution of comonomer units; (b) copolymers containing blocks of vinyl monomer (carrying epoxy or hydroxyl group) separated by sequences of randomly distributed comonomers; (c) copolymers containing random distribution of both monomers. These facts have important implications for use of these copolymers for preparing antifouling coatings of adjustable mechanical properties. Since the homopolymers of these organotin monomers are elastomeric (1), it follows that copolymers containing blocks of organotin monomer units, such as copolymer of tri-n-butyltin methacrylate with N-methylolacrylamide, should give a polymer with improved impact strength. Similarly, the copolymers containing blocks of epoxy or hydroxyl group -containing monomer should give polymers with improved tensile strength, while the copolymers containing random distributions of two comonomers might give the slowest leaching rate. Thus the range of useful properties that can be varied by using these copolymers, which provide a means for controlling both the concentration and distribution of crosslinking sites, is broad.

B.   Crosslinking Reactions

A 46:54 copolymer of tri-n-butyltin methacrylate with glycidyl methacrylate was chosen for all copolymer crosslinking reactions. As discussed above, this composition corresponds to azeotropic copolymerization of the two monomers which yields a homogeneous copolymer regardless of conversion.

Two aliphatic amines, viz. diethylenetriamine (DETA) and triethylenetetramine (TETA), and two aromatic amines, namely m-phenylene diamine (MPDA) and 4,4´-methylene dianiline (MDA), were used as crosslinking agents. Various amounts of amines corresponding to 75%, 85%, 100%, and 110% of stoichiometrically required quantities were used. The crosslinked product was inspected for tackiness, hardness and toughness and then extracted with benzene to remove uncrosslinked portions. The results are summarized in Table I. From this table it can be seen that the amount of extractables is much larger for the cases where aliphatic amines are used as curing agents. Since aliphatic amines react much more rapidly than aromatic amines with epoxy groups, it is probable that in the former case, the curing agent is used up inefficiently to form intramolecular bonds, thus leaving large portions of linear polymers uncrosslinked. It is also seen that the optimum level of hardness and toughness is achieved when the concentration of crosslinking agent is between 75% and 85% of the theoretically required amount.

It was interesting to find that the copolymer of tri-n-butyltin methacrylate with glycidyl methacrylate could be

Table I. Extractables and Tin Content of Crosslinked Azeotropic Copolymer of Tri-n-butyltin Methacrylate with Glycidyl Methacrylate

Crosslinking Agent	Concentration of Crosslinking Agent(a)	Extractables %	Tin Content After Extraction % by Weight	Tin Content Before Extraction % by Weight	Characteristics Estimated by Inspection
DETA	75	18.84	16.59	21.01	hard, non-tacky, tough
	85	19.13	16.26	20.81	hard, non-tacky, tough
	100	20.75	14.89	20.62	non-tacky, very brittle, hard
	110	22.61	14.48	20.43	non-tacky, very brittle, hard
TETA	75	17.60	15.55	20.81	hard, non-tacky, tough
	85	18.30	16.06	20.62	hard, non-tacky, tough
	100	18.03	15.84	20.43	non-tacky, very brittle, hard
	110	18.36	14.59	20.25	non-tacky, very brittle, hard
MDA	75	8.47	17.49	19.36	hard, tough, non-tacky
	85	8.46	17.04	19.04	hard, very tough, non-tacky
	100	7.63	16.86	18.71	hard and brittle, non-tacky, cracks on bending
	110	7.59	16.99	18.40	hard and very brittle, non-tacky, cracks on bending
MPDA	75	12.67	19.55	20.62	hard, tough, non-tacky
	85	11.83	18.96	20.43	hard, very tough, non-tacky
	100	11.36	18.40	20.25	hard and brittle, non-tacky, cracks on bending
	110	10.82	17.69	19.88	hard and very brittle, non-tacky, cracks on bending

(a) Percent of theoretical stoichiometric amount required.

self-cured to a crosslinked product simply by heating it.   The extractables for this case were found to be 12.2%.   The crosslinking is presumably through epoxy groups catalyzed by tri-n-butyltin ester groups.   Similarly, the copolymer could also be crosslinked using uranyl nitrate as catalyst and, in this case, the extractables were 12.7%.

The typical structure of the network polymer obtained by reaction of epoxy-group-containing resins with amines is shown in Fig. 1.   It should be noted that the structure has a

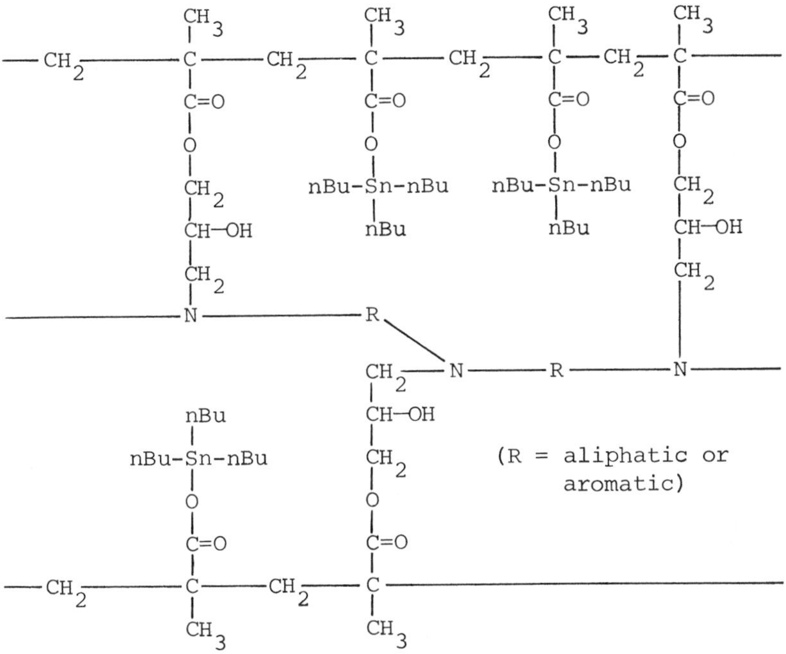

Fig. 1.   Typical structure of network polymer obtained by reaction of epoxy-group-containing copolymers with amines.

preponderance of hydroxyl groups in it.   Besides the fact that amines catalyze hydrolysis of esters of carboxylic acids, it is also known that neighboring hydroxyl groups can catalyze ester hydrolysis.   Thus the network structure shown in Fig. 1 would lead to higher leaching rates by increasing the rate of hydrolysis.   The relationship of such structural features to biocidal activity of the polymers needs careful investigation.

In contrast, the homopolymerization of epoxy groups to polyethers, which is the cause of crosslinking by self-curing, would not lead to the formation of a large number of hydroxyl groups in the network structure.   Catalysis of ester hydrolysis by hydroxyl groups would not, therefore, be expected in the self-cured system.

Although the crosslinking reactions discussed here show
how to obtain thermoset organotin epoxy-based polymers, it
should be apparent that thermoset polyurethane-based organotin
polymers can also be obtained by reacting the hydroxyl-group-
containing organotin copolymers, such as the copolymer of tri-
n-butyltin methacrylate with N-methylolacrylamide, with a
diisocyanate. Such copolymers have been prepared and are
being investigated in detail.

C.   Biotoxicity of Crosslinked Polymers

A modified petri-dish method was used to test biotoxicity
of the crosslinked polymers. Since the crosslinked polymers
are insoluble, it was necessary to make a small cylindrical
hole in the agar medium and fill this hole with the powdered
polymer to be tested. The petri dishes were next incubated
for a period of time, then developed, and the inhibition zone
measured.

The crosslinked polymers were tested for toxicity to
three microorganisms, viz. Pseudomonas nigrifaciens (marine
bacterium), Glomerella cingulata (soil fungus), and Sarcina
lutea (soil bacterium). Figure 2 shows the typical zones of
inhibition of Glomerella cingulata caused by various crosslinked and linear organotin polymers. Since the particle size
and hence the surface area of all test samples of polymers
were approximately the same, the differences in leaching rate
of the toxin can be detected by the width of the annular inhibition zones. Similar data were obtained when using other
test organisms. All these results are summarized in Table II.
This table shows that the copolymer cured with aromatic amines
gave inhibition zones ranging from 1 to 6 mm, while the copolymer crosslinked with aliphatic amines gave inhibition zones
ranging from 6 to 14 mm. Since the base copolymer used in all
crosslinking reactions was the same, the differences in leaching rate of toxin and hence the biocidal action of crosslinked
polymers are directly attributable to the degree of crosslinking and the nature of the crosslinking agent. The results
suggest that increasing the degree of crosslinking and stiff-
ness of the curing agent lead to reduced leaching rate of
toxin and hence reduced inhibition zones. Thus, the biotoxicity of crosslinked organometallic polymers can be controlled
by controlling the degree of crosslinking and the nature of
crosslinks, to obtain optimum biocidal action.

IV.   SUMMARY AND CONCLUSIONS

Tri-n-butyltin acrylate and tri-n-butyltin methacrylate
were copolymerized with a number of vinyl monomers containing

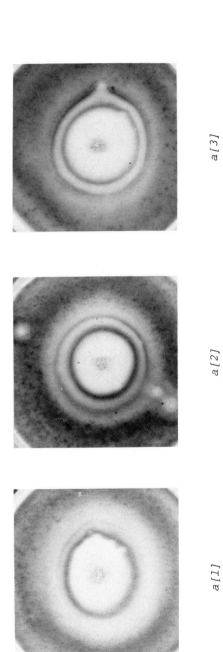

a[1]

a[2]

a[3]

Fig. 2. Inhibition of *Glomerella cingulata* growth by *poly(tri-n-butyltin meth-acrylate-co-glycidyl methacrylate) crosslinked by (a) DETA; (b) MDA; (c) uncrosslinked copolymer. The incubation times were [1] 24 hours; [2] 48 hours; [3] 96 hours.*

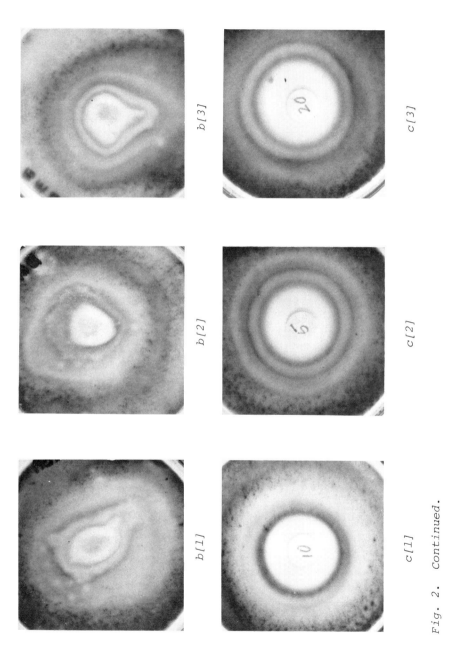

b[3]

c[3]

b[2]

c[2]

b[1]

c[1]

Fig. 2. Continued.

192

Table II: Inhibition of Test Microorganisms by Crosslinked and Linear Poly(tri-n-butyltin methacrylate-co-glycidyl methacrylate)

Crosslinking Agent	Test Organism (1)			Test Organism (2)			Test Organism (3)		
	(a) hr	(b) mg	(c) mm	(a) hr	(b) mg	(c) mm	(a) hr	(b) mg	(c) mm
110[(d)]-TETA	24	6.2	9	15	5.2	7	15	5.0	7
	48	7.7	6	24	4.3	7	24	5.3	10
	72	5.9	10	39	5.6	9	39	4.3	10
	96	5.0	7	48	7.0	11	48	4.5	10
110-DETA	24	4.4	9	15	4.7	9	15	8.0	10
	48	6.5	9	24	4.6	7	24	5.0	10
	72	5.2	14	39	6.2	7	39	4.6	10
	96	5.5	6	48	4.4	9	48	7.9	11
85-MDA	24	6.4	1	15	6.5	5	15	5.5	5
	48	5.5	1	24	7.1	4	24	6.5	4
	72	5.0	4	39	5.4	2	39	5.3	4
	96	5.1	4	48	7.4	4	48	6.5	4
75-MPDA				15	6.2	4	15	4.8	4
				24	4.1	5	24	6.3	5
				39	5.6	4	39	5.4	5
				48	6.4	6	48	6.0	4
None (Linear copolymer)	15	5.0	14	15	5.0	8	15	5.0	16
	24	5.0	16	24	5.0	9	24	5.0	17
	39	5.0	17	39	5.0	9	39	5.0	20
				48	5.0	9	48	5.0	17

Test Organisms:

   (1) Pseudomonas nigrifaciens, a marine bacterium
   (2) Sarcina lutea, a soil bacterium
   (3) Glomerella cingulata, a soil fungus

[a] Incubation time, hours

[b] Weight of crosslinked or linear product used, mg

[c] Inhibition zone size, mm

[d] Percent of theoretical stoichiometric amount required

an epoxy or a hydroxyl group, by solution polymerization method using free-radical initiators as catalysts. The crosslink-
ability of these copolymers was demonstrated by crosslinking a
46:54 copolymer of tri-n-butyltin methacrylate with glycidyl
methacrylate, with a number of aliphatic and aromatic amine
curing agents. This copolymer could also be cured with uranyl
nitrate. Finally, it was found that this copolymer could be
self-cured simply by heating without added catalysts or curing
agents. All crosslinked polymers were inspected for physical
properties and their sol fractions determined by extractions
with benzene.

The biotoxicity data on the crosslinked polymers prepared
above were obtained, using the culture-plate method, by studying the inhibition of three test organisms, viz. Pseudomonas
nigrifaciens (marine bacterium), Sarcina lutea (soil bacterium), and Glomerella cingulata (soil fungus). It was shown
that the degree and the nature of crosslinking have a significant effect on the size of inhibition zones. When aromatic
amines, which give stiff crosslinks, were used as curing
agents, the crosslinked product gave the smallest inhibition
zones and hence the lowest leaching rate of toxin among all
crosslinked polymers tested. Thus the degree and the nature
of crosslinks can be varied to optimize the biocidal action of
thermoset organotin polymers.

V.   ACKNOWLEDGEMENTS

Grateful acknowledgement is made to the David W. Taylor
Ship Research and Development Center, Annapolis, for the sup-
ply of Pseudomonas nigrifaciens and partial support through a
contract.

VI.   REFERENCES

1. Subramanian, R.V., Garg, B.K., Jakubowski, J., Corredor,
    J., Montemarano, J.A., and Fischer, E.C., 172nd Natl.
    Meeting, Am. Chem. Soc., Div. Org. Coat. Plast. Chem.,--
    Prepr., 36(2), 660 (1976).
2. Subramanian, R.V., and Anand, M., 172nd Natl. Meeting,
    Am. Chem. Soc., Div. Org. Coat. Plast. Chem., Prepr.,
    36(2), 233 (1976); also in "Chemistry and Properties
    of Crosslinked Polymers" (S. Labana, Ed.), p. 1, Aca-
    demic Press, New York, 1977.
3. Montermoso, J.C., Andrews, T.M., and Marinelli, L.P.,
    J. Polym. Sci., 32, 523 (1958).
4. Gilman, H., and Rosenberg, D., J. Am. Chem. Soc., 75, 3592
    (1953).
5. Garg, B.K., Corredor, J., and Subramanian, R.V., J. Macro-
    mol. Sci.-Chem., in press.

INFRARED AND NUCLEAR MAGNETIC RESONANCE
ANALYSIS OF ORGANOTIN TOXICANTS FOR
MARINE ANTIFOULING COATINGS

James F. Hoffman, Keith C. Kappel,
Lydia M. Frenzel and Mary L. Good
University of New Orleans

Infrared and nuclear magnetic resonance spectroscopies
have been employed to elucidate the chemical and structural properties of triorganotin compounds incorporated
into marine antifoulant coating formulations. It is
necessary to determine the molecular species of the
toxicant in order to be able to predict the mechanism
of toxicant release and to design new, more effective
coatings. In two conventional vinyl antifoulant coatings, the toxicant was found to react with one component in the coating formulation to yield different
triorganotin compounds. A change in the $\nu_{asym}(SnC)$
vibration was found to correlate with a change in the
coordination geometry of tin. These results indicate
that infrared spectroscopy is a powerful technique
that can provide insights into the chemical processes
occurring in the antifouling coatings.

I. INTRODUCTION

The problem of marine "fouling" of ship hulls and marine
installations is well known and has been attacked by many research and development groups. Even the popular press and
general technology periodicals have recently highlighted the
problem in their publications (1,2,3). The present national
difficulties associated with energy shortages make this
fouling problem even more crucial. For example, "the drag
caused by only a six-month accumulation of marine foulants
can force a vessel to burn 40 percent more fuel just to maintain normal cruising speed" (1). Thus any attempt at fuel
economy in the shipping industry and in the Navy must place
the development of effective antifoulant procedures as a high
priority item. The only practical solution through the years
has been the utilization of paints and coatings having antifoulant activity. Most of the coatings which have been

developed are paint formulations containing a toxicant component. The most widely used toxicant has been cuprous oxide
although various compounds of arsenic, mercury, lead and tin
have been used. The cuprous oxide coatings are good fouling
inhibitors but they have rather short effective life-times.
The arsenic, mercury and lead coatings have severe toxicological problems and have been either restricted in use or completely removed from the commercial market. The tin containing coatings have been investigated by a number of groups,
particularly by researchers at the Naval Ship Research and
Development Center at Annapolis, Maryland. Studies have
shown that tin antifoulants possess several advantages over
conventional copper **antifoulants** (4,5). For example, they
provide superior **antifouling** over a longer period of time.
When applied on conductive substrates, they do not promote
corrosion as do copper based coatings. They exhibit control
over several types of algae and barnacles when properly formulated. It has been proposed that the organotin compounds
degrade to non-toxic, non-cummulative inorganic forms of tin,
but there is no experimental evidence to support this propo-
sition (6,7,8). Thus, the most promising antifoulant coatings
which have been designed and characterized are those containing organotin toxicants.

Although the formulation of new coatings containing
organotin moieties has been an area of active research, the
determination of the chemical form of the toxicant in the
bulk and on the surface of the coatings, and the elucidation
of the surface release mechanism have received less attention. Thus, this initial report on our studies of these
coating systems will review our efforts to devise procedures
and background data which will allow the nature of the organotin toxicants to be determined by infrared and proton NMR
spectroscopies. It is anticipated that these studies and
other related work will eventually provide the tools for the
identification of the chemical composition of the tin toxicants as a function of the coating matrix, the curing process,
actual aging in normal use, and coating efficiency. Information of this type is essential for the design of new, more
effective coatings and for the evaluation of the long term
environmental impact of the widespread use of these toxic
material.

The most effective and widely used organotin coating
additives are the triorganotin compounds (R_3SnX where
R = alkyl or phenyl and X = F, Cl, Br, I, OH, OCOR, $OSnR_3$,
SnR_3 or acrylate). The tributyltin compounds have been used
most extensively to date. These materials exhibit a much
higher level of toxicity to marine foulants than they do to
mammals. The shorter chained alkyls exhibit a high toxicity
to mammals and a low toxicity to marine life (9). Thus this
initial study is devoted to the evaluation of coatings containing trialkyltin species with special emphasis on the

tributyl series. Two general classes of coatings are being
investigated: (1) conventional formulations consisting of an
organotin compound, usually either bis(tri-n-butyltin) oxide
(TBTO) or tri-n-butyltin fluoride (TBTF), thoroughly mixed
with the rest of the coating components; and (2) newly developed coating systems consisting of R_3SnX moieties (where X is
vinyl, acrylate, or maleate) polymerized or copolymerized
through the unsaturated X groups. The conventional coatings
presumably have free toxicant not bound to the coating matrix
while in the polymer materials the toxicant species is covalently bonded to the matrix. Neat organotin materials have
been examined to provide baseline spectra and spectral assignments of functional groups. Correlation techniques have been
used to assign the SnC_3, Sn-X, Sn-0 and COO vibrational modes
in the infrared spectra. These assignments have then been
used for the identification of functional groups and chemical
entities in the coating formulations.

II. EXPERIMENTAL

     Transmission infrared spectra for liquid samples were ob-
tained as thin films between KBr and polyethylene disks in the
region 4000-200 cm^{-1}. The spectra of solid samples were
obtained as KBr pellets. The spectra were recorded on a
Perkin-Elmer Infrared Spectrophotometer Model 283 equipped
with a printer accessory, which provides a digital readout of
peak positions and relative intensities of the peaks. Peak
positions are accurate to **+3 cm**$^{-1}$ **in** the region 4000-2000
cm^{-1}, and to ±1.5 cm^{-1} in the region 2000-200 cm^{-1}. Proton
nuclear magnetic resonance (^1H nmr) spectra were recorded on
a Hitachi Perkin-Elmer R-20B spectrometer operating at a frequency of 60.0 MHz. The organotin and polymer samples were
dissolved in $CHCl_3$, $CDCl_3$, and CCl_4 to give a 10% (w/v) solution with a few drops of tetramethylsilane (TMS) added as an
internal reference.
     Organotin compounds were obtained from M&T Chemicals,
Inc. and were used without further purification. Purity was
checked by chemical analysis and by ^1H nmr. The conventional
coatings, which contained $(Bu_3Sn)_2O$ and Bu_3SnF were donated by
Glidden-Durkee, Standard Paint and Varnish Company of New
Orleans, and by the U. S. Navy. All coatings investigated
were vinyl based formulations. The Glidden-Durkee coating
contains $(Bu_3Sn)_2O$ as the toxicant. The coating from the
Standard Paint and Varnish Company (Alum-A-Tox) contains
Bu_3SnF. The Navy formulation, 1020A, contains both $(Bu_3Sn)_2O$
and Bu_3SnF as toxicants. Formulations were also obtained
which did not contain the toxicants. Experimental organotin
containing polymers (OMP's) were donated by Dr. E. Fischer of
the David Taylor Naval Ship Research and Development Center
(DTNSRDC). The compositions of these polymers are as follows:

OMP-1 is poly(tri-n-butyltin methacrylate/tri-n-propyltin methacrylate/methyl methacrylate); OMP-2 is poly(tri-n-butyltin methacrylate/methyl methacrylate); OMP-4 is the tri-n-butyltin ester of poly(methylvinylether/maleic acid). The formulas for these materials are shown diagramatically below.

$$\left[CH_2-\underset{\underset{\underset{O-CH_3}{|}}{\underset{C=O}{|}}{\overset{\overset{CH_3}{|}}{C}}\right]_x \left[CH_2-\underset{\underset{\underset{O-SnBu_3}{|}}{\underset{C=O}{|}}{\overset{\overset{CH_3}{|}}{C}}\right]_y \quad and \quad \left[CH_2-\underset{\underset{\underset{O-CH_3}{|}}{\underset{C=O}{|}}{\overset{\overset{CH_3}{|}}{C}}\right]_u \left[CH_2-\underset{\underset{\underset{O-SnPr_3}{|}}{\underset{C=O}{|}}{\overset{\overset{CH_3}{|}}{C}}\right]_z$$

OMP 1

$$\left[CH_2-\underset{\underset{\underset{O-CH_3}{|}}{\underset{C=O}{|}}{\overset{\overset{CH_3}{|}}{C}}\right]_x \left[CH_2-\underset{\underset{\underset{O-SnBu_3}{|}}{\underset{C=O}{|}}{\overset{\overset{CH_3}{|}}{C}}\right]_y \qquad \left[CH_2-\underset{\underset{Bu_3Sn-O}{|}}{\underset{C=O}{|}}{\overset{\overset{OCH_3}{|}}{CH}}-\underset{\underset{O-SnBu_3}{|}}{\underset{C=O}{|}}{CH}-CH\right]_x$$

OMP 2   OMP 4

III.   RESULTS AND DISCUSSION

Assignments of the infrared vibrations of the triorganotin compounds were made by the correlation method, since these are complex molecules and few spectral assignments of comparable compounds are available in the literature (10,11). The Sn-X, Sn-O, and SnC_3 vibrations are of particular interest because these functional groups are the most likely ones to be affected by the leaching of the toxicant from the coating matrix. All of these vibrations occur below 1100 cm^{-1}. Other vibrations of importance to this study are ν_{OH}, which occurs around 3600 cm^{-1}, and the asymmetric and symmetric stretching frequencies, $\nu_{asym(COO)}$ and $\nu_{sym(COO)}$, of the carboxylate group. These vibrations are located between 1650 and 1300 cm^{-1}. Assignments of the vibrational modes of interest are presented in Table I.

Triorganotin compounds (R_3SnX) exhibit either a tetrahedral or a trigonal bipyramidal geometry depending on the nature of the X group. When X is capable of acting as a bridging ligand (e.g. F, OH, $OCOCH_3$), the organotin compound exhibits a trigonal bipyramidal geometry with the X group bridged between planar SnR_3 moieties. With other X groups,

TABLE I

Assignments of Infrared Bands

Compound	ν_{Sn-X}	$\nu_{asym}(Sn-O-Sn)$	$\nu_{asym}(SnC)$	$\nu_{sym}(SnC)$	$\nu_{asym}(COO)$
Pr$_3$SnCl	330		593	507	
Bu$_3$SnCl	323		597	509	
Ph$_3$SnCl	330				
Bu$_3$SnF	328		612	510	
Ph$_3$SnF	368				
Bu$_3$SnBr	230		596	505	
Bu$_3$SnI	–		593	503	
(Pr$_3$Sn)$_2$O		773	586	506	
(Bu$_3$Sn)$_2$O		770	591	505	
(Ph$_3$Sn)$_2$O		772			
(Bu$_3$Sn)$_2$			584	495	
Bu$_3$SnOAc			609	489	1572
Bu$_3$Sn acrylate			607	508	1536
Bu$_3$Sn methacrylate			595	509	1618

the tin atom is four coordinate with a distorted tetrahedral geometry. Due to the complexity of these triorganotin molecules, a composite set of vibrations (asymmetric and symmetric stretching frequencies) will be observed for the SnC_3 part of the molecule rather than a single Sn-C vibration. For trigonal bipyramidal molecules with D_{3h} site symmetry only the $\nu_{asym(SnC_2)}$ is infrared active. For the molecules with C_{3v} site symmetry (distorted tetrahedron) the $\nu_{asym(SnC)}$ and $\nu_{sym(SnC)}$ are both infrared active. Experimentally, two vibrations are observed for all of the trialkyltin compounds in the SnC_3 region. The infrared inactive $\nu_{sym(SnC)}$ mode is a weak band in the infrared for the five coordinate complexes which indicates that the site symmetry of the tin compounds is lower than D_{3h}. The $\nu_{asym(SnC)}$ is sensitive to changes in coordination number as is shown in Table II. Five coordinate

TABLE II.

$\nu_{asym(SnC)}$ vs. Coordination Number of Tin

Compound	$\nu_{(cm^{-1})}$	Coordination Number
$(Bu_3Sn)_2$	584	4
Bu_3SnCl	596	4
Bu_3SnBr	596	4
Bu_3SnI	593	4
$(Bu_3Sn)_2O$	591	4
Pr_3SnCl	593	4
$(Pr_3Sn)_2O$	586	4
Bu_3Sn methacrylate	595	4
Bu_3SnF	612	5
Bu_3SnOAc	610	5
Bu_3Sn acrylate	607	5

triorganotin compounds exhibit the $\nu_{asym(SnC)}$ between 603-612 cm^{-1}; whereas, the $\nu_{asym(SnC)}$ occurs between 584-597 cm^{-1} for the four coordinate complexes. The asymmetric SnC_3 stretching frequency is observed in the organotin containing polymers (OMP's), and occurs between 587-592 cm^{-1}. This implies that the tin moiety is four coordinate in these formulations. For the conventional coatings investigated in this study, the $\nu_{asym(SnC)}$ is unobservable because of vibrations which occur in this region of the spectrum from other paint components. The Sn-O-Sn asymmetric vibration, $\nu_{asym(SnOSn)}$, has been

assigned to an intense band between 770-775 cm^{-1} for molecules
of the type $(R_3Sn)_2O$ (see Table I).   In the commercially
available coating formulations which contain bis(tri-n-
butyltin) oxide (TBTO) as the toxicant (i.e. Glidden-Durkee
and Navy 1020A), the ν_{asym}(SnOSn) is not observed, even though
the concentration of toxicant is high enough to be easily observed by infrared analysis.   In the Navy 1020A formulation,
which also contains tri-n-butyltin fluoride (TBTF) as toxicant, the Sn-F vibration is observed.   The toxicant is presumably unchanged upon addition to the coating formulation.
In the two coatings which employ TBTO as the toxicant, a band
is observed at 1634 cm^{-1} in the spectrum of the Glidden-Durkee
formulation, and at 1648 cm^{-1} in the spectrum of the Navy
1020A formulation.   These bands were not observed in either
TBTO, TBTF, or the paint formulations without toxicant.   The
absence of ν_{asym}(SnOSn) and the presence of the band around
1640 cm^{-1} implies that the toxicant (TBTO) undergoes a
reaction upon mixing with one of the components of the paint.
The peak around 1640 cm^{-1} can be assigned with considerable
certainty to the formulation of a tributyltin carboxylate in
both of these coatings.

     Upon addition of TBTO to both the Navy 1020A and Glidden-
Durkee formulations without toxicant, an infrared spectrum
identical to the spectrum of the formulation with toxicant was
observed.   (see Figure 1B and 1D, and Figure 2B and 2C).   No
ν_{asym}(SnOSn) was observed in these spectra which again indicates that TBTO has undergone a reaction.   For the Navy 1020A
formulation without toxicant, the addition of tributyltin acetate (TBTA) to this coating yields a material which has an
identical spectrum to that in Figure 1B.   When TBTF is added
to the formulation without toxicant, no reaction is observed
(see Figure 1C).

     It has been reported that bis(triethyltin) oxide reacts
with carboxylic acids and esters (12).   The TBTO in Navy 1020A
was found to react with butyl acetate, one of the components
of the solvent, according to the following reaction:

$$(Bu_3Sn)_2O + 2CH_3\overset{O}{\overset{\|}{-C}}-O-CH_3 \longrightarrow 2CH_3\overset{O}{\overset{\|}{-C}}-O-SnBu_3 + (C_4H_9)_2O$$

Tributyltin acetate precipitates from the reaction mixture
upon cooling.

     The ν_{asym}(COO) vibration in TBTA in the solid state
occurs at 1572 cm^{-1}, but when dissolved in an inert solvent
such as chloroform, the band is shifted to 1640 cm^{-1}.   This
finding is in complete agreement with our assignment of the
1640 cm^{-1} band of the TBTO reaction product to the formation
of TBTA in the coating matrix.

     The Glidden-Durkee coating uses methyl isobutyl ketone
and xylenes as a solvent system, and TBTO does not react with
either component.   However, this formulation also contains
rosin, a natural product which is primarily abietic acid,

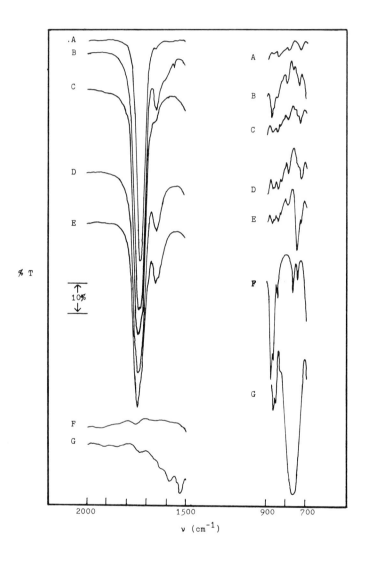

Figure 1.   Infrared spectra of Navy 1020A coatings and triorganotin toxicants.   (A) without toxicant, (B) with toxicant, (C) without toxicant plus added TBTF, (D) without toxicant plus added TBTO, (E) without toxicant plus added TBTA, (F) neat TBTF, (G) neat TBTO.

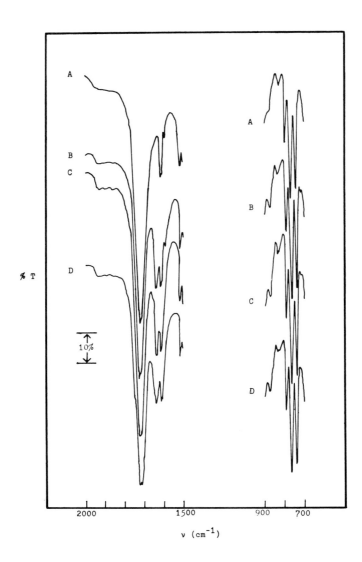

Figure 2.   Infrared spectra of Glidden-Durkee coatings.
(A) without toxicant, (B) with toxicant, (C) without toxicant
plus added TBTA, (D) without toxicant plus added TBTO.

$C_{19}H_{29}COOH$. Bis(tributyltin) oxide does react with the rosin to yield the tributyltin ester of abietic acid, according to the following reaction (13):

$$(Bu_3Sn)_2O + 2C_{19}H_{29}COOH \longrightarrow 2C_{19}H_{29}-\overset{O}{\overset{\|}{C}}-O-SnBu_3 + H_2O$$

The infrared spectrum of the reaction mixture exhibits a band at 1637 cm^{-1} in the spectrum of the Glidden-Durkee coating with toxicant. The products of the above reaction were not isolated.

The Alum-A-Tox coating system contains TBTF as the toxicant. It appears that no reaction occurs between the toxicant and the paint formulation, but a conclusive investigation by infrared is not possible due to bands of the paint formulation which overlap the vibrations of interest. Tributyltin fluoride did not react with the Navy 1020A so it appears likely that it will not react with the Alum-A-Tox.

The information obtained and conclusions reached on the reactions of the toxicants in the antifoulant formulations is in agreement with the observations obtained by Mossbauer spectroscopy, which are presented in the following paper.

IV.   CONCLUSIONS

From this investigation, it is apparent that the toxicant in the coating must be identified in order to accurately predict the lifetime, effectiveness and environmental impact of antifouling coatings. Bis(tributyltin) oxide which was added as the toxicant in two of the conventional formulations, was found to react with one component of the paint formulation to yield a tributyltin carboxylate. Infrared and nuclear magnetic resonance spectroscopies were found to be excellent tools for the identification of the organotin species and for the investigation of molecular rearrangements.

Future work will be directed to the identification of changes in the toxicant during and after exposure of the coating to aqueous and saline environments. Efforts will also be expended to determine the chemical form of the organotin species leached from the coatings into the environment.

V.   ACKNOWLEDGEMENTS

This work is a result of research sponsored by the Office of Naval Research and the NOAA Office of Sea Grant, Department of Commerce, under Grant #R/MTR-1. We also acknowledge the generous gift of the OMP's by Dr. E. Fischer of the D. Taylor Naval Ship Research and Development Center. One of us (L.M.F.) wishes to thank Charles A. Frenzel of Coastal Science Associates for his helpful discussions.

VI. REFERENCES

1. Starbird, E. A., and Sisson, R. F., National Geographic
    November, 623 (1973).
2. Starbird, E. A., The Readers Digest March, 123 (1974).
3. Murray, C., Chemical and Engineering News January 6, 18
    (1975).
4. Vizgirda, R. J., Paint and Varnish Prod. 62, No. 12, 25
    (1972).
5. Evans, C. J., J. Paint Tech. 34, Aug., 17 (1970).
6. Cremer, J. E., Biochem. J. 67, 87 (1957).
7. Sheldon, A. W., J. Paint Tech. 47, Jan., 54 (1975).
8. Smith, P., and Smith, L., Chem. in Britain 11, 208
    (1975).
9. Evans, C. J., and Smith, P. J., J. Oil Col. Chem. Assoc.
    58, 160 (1975).
10. Mendelsohn, J., Marchand, A., and Valade, J.,
    J. Organomet. Chem. 6, 25 (1966).
11. Geissler, H., and Kriegsmann, H., J. Organomet. Chem. 11,
    85 (1968).
12. Anderson, H. H., J. Org. Chem. 19, 1766 (1954).
13. Anderson, H. H., J. Org. Chem. 22, 147 (1957).

DETERMINATION OF ORGANOTIN STRUCTURES IN ANTIFOULING COATINGS BY MÖSSBAUER SPECTROSCOPIC TECHNIQUES

Elmer J. O'Brien, Charles P. Monaghan and Mary L. Good
University of New Orleans

In characterizing the leaching properties of antifouling coatings containing organotin compounds, it has become necessary to determine the structure of the tin species in the coating. An extensive investigation of the Mössbauer parameters of numerous tin compounds of the type R_3SnX (R = n-butyl, n-propyl, phenyl; X = I, Br, Cl, OH, OAc, F) has shown that the quadrupole splitting parameter and ρ (ratio of isomer shift to quadrupole splitting) are related to the molecular geometry. The Mössbauer parameters for organotin polymers used in antifouling coatings and conventional coating formulations containing organotin compounds have been obtained. The quadrupole splitting and ρ value in these cases are sensitive to structural changes which occur in the organotin compounds upon incorporation into these coating matrixes. Mössbauer studies have been performed on the compounds $[(C_4H_9)_3Sn]_2O$, $(C_4H_9)_3SnOAc$, $(C_4H_9)_3SnF$ in various matrices and solvent systems, and the results are discussed in the context of the behavior of these compounds in the coating systems.

I. INTRODUCTION

Certain organotin compounds of the type R_3SnX (R = alkyl, phenyl; X = Cl, F, OAc, etc.) have been employed in coating formulations as marine antifouling additives. The effectiveness of the coating is dependent upon the identity of the tin species, the concentration of the solvated tin species at the coating surface, and the leach rate of the tin species out of the coating. In many coatings the organotin compounds are added directly to the basic paint formulations, but recently polymers containing organotin pendant groups have been synthesized and tested for antifouling effectiveness. To fully characterize the role of the organotin components, the identity and structure of the tin species in the various coatings must be ascertained.

The quadrupole splitting parameter in Mössbauer spectroscopy is sensitive to the geometry of the electronic field
about the tin moiety, and the isomer shift parameter is indicative of the electron density at the tin nucleus which
assists in assigning the oxidation state of the tin (1). Some
researchers have also used the parameter ρ, the ratio of the
quadrupole splitting to the isomer shift (relative to SnO_2),
as an empirical relationship indicative of the coordination
number of tin(IV) compounds (2,3). A ρ value greater than 2.1
is considered to be indicative of a coordination greater than
four; whereas, a ρ value less than 1.8 is considered to be
indicative of four-fold coordination. This relationship is
based on the observation that the isomer shift values generally decrease and that the quadrupole splitting values increase
with coordination numbers greater than four. By examining ρ
values the geometry of the organotin species in a coating can
be deduced.

The Mössbauer data in the literature are characterized
by discrepancies from author to author (4). Many of the
earlier measurements were taken with crude instrumentation
on compounds of questionable purity. Variations in data due
to crude instrumentation are to be expected but the effect of
a compound's purity is more subtle. Figure 1 shows a Mössbauer spectrum of impure $[(C_4H_9)_3Sn]_2O$. The spectrum was
first interpreted as two lines giving an isomer shift of
1.25mm/sec and a quadrupole splitting of 1.64mm/seC. The compound was determined to be impure and the spectrum was eventually fit to four lines, indicating the presence of two compounds. After purifying $[(C_4H_9)_3Sn]_2O$ by vacuum distillation
and taking another Mössbauer spectrum, the new measurement
yielded an isomer shift of 1.26mm/sec and a quadrupole splitting of 1.49mm/sec. To provide acceptable working standards
for determining the chemical and physical parameters of tin
moieties in the coatings of interest, it has been necessary
to redetermine the Mössbauer parameters for a wide variety of
carefully characterized organotin compounds. This report contains our initial studies on the applicability of Mössbauer
spectroscopy as a tool in determining the chemical characteristics and solid state structures of the organotin species
used in marine antifouling coatings.

II.   EXPERIMENTAL

The ^{119}Sn Mössbauer data were obtained on an Austin
Science Associates constant acceleration spectrometer operated
in conjunction with a Nuclear Data 2200 multichannel analyzer.
The source was $Ba^{119}SnO_3$, obtained from New England Nuclear
Corporation, at an activity of 10 mCi. The data was collected
with the source at room temperature and the absorber at 77°K.
No correction has been made for the second order Doppler

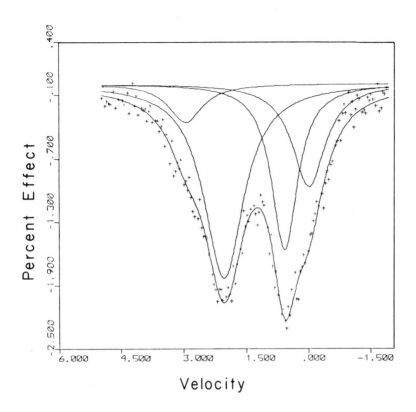

Figure 1. Mössbauer spectrum of impure $[(C_4H_9)_3Sn]_2O$.

effect. Velocity calibration was accomplished with an Austin Science Associates laser interferometry system. Spectra were corrected for the displacement of the source from equilibrium position and then fitted to a sum of Lorentzian curves by a least squares procedure. All isomer shift values are refer- enced to $BaSnO_3$ at room temperature. The results for R_3SnX compounds are presented in Table I, and the results for orga- notin species in various coatings are presented in Table II. All reported parameters are the weighted averages obtained from at least two spectra.

The organotin compounds were acquired from M&T Chemicals, Inc. and were used as received. All organotin compounds have been analyzed for C, H, and Sn. The formulations presented in Table I are consistent with the results obtained.

Frozen solution samples were prepared by placing a 10% solution in a liquid cell and quick freezing in liquid nitro-

TABLE I

Mössbauer Parameters for Various
Organometallic Tin Compounds*

Compound	Isomer Shift (mm/sec)	Quadrupole Splitting (mm/sec)	ρ
$(C_3H_7)_3SnCl$	1.563+0.004	3.438+0.009	2.228
$(C_3H_7)_3SnBr$	1.51 +0.02	3.24 +0.04	2.14
$(C_3H_7)SnI$	1.57 +0.01	2.89 +0.02	1.84
$[(C_3H_7)Sn]_2O$	1.28 +0.02	1.50 +0.03	1.17
$(C_4H_9)_3SnF$	1.422+0.002	3.871+0.006	2.722
$(C_4H_9)_3SnCl$	1.58 +0.01	3.49 +0.01	2.21
$(C_4H_9)_3SnBr$	1.56 +0.04	3.32 +0.06	2.13
$(C_4H_9)_3SnI$	1.56 +0.02	2.88 +0.03	1.85
$(C_4H_9)_3SnOAc$	1.439+0.003	3.531+0.005	2.454
$[(C_4H_9)_3Sn]_2CO_3$	1.46 +0.06	3.65 +0.08	2.50
$[(C_4H_9)_3Sn]_2O$	1.256+0.003	1.490+0.008	1.186
$(C_4H_9)_3SnOMacr$	1.458+0.005	3.62 +0.01	2.48
$(C_4H_9)_3SnOAcr$	1.44 +0.01	3.61 +0.02	2.51
Ph_3SnF	1.257+0.005	3.65 +0.01	2.90
Ph_3SnCl	1.35 +0.01	2.54 +0.01	1.88
Ph_3SnBr	1.40 +0.01	2.45 +0.01	1.75
Ph_3SnI	1.45 +0.01	2.24 +0.01	1.54
Ph_3SnOAc	1.32 +0.01	3.35 +0.01	2.54
Ph_3SnOH	1.19 +0.01	2.84 +0.01	2.49
$[Ph_3Sn]_2O$	1.16 +0.01	1.38 +0.01	1.19

*Abbreviations: -OAc is $-O(CO)CH_3$, -OMacr is $-O(CO)C(CH_3)CH_2$,
-OAcr is $-O(CO)CHCH_2$, and Ph is C_6H_5.

TABLE II

Mössbauer Results on Coating Systems*

Coating Type	Code	Isomer Shift (mm/sec)	Quadrupole Splitting (mm/sec)	ρ	Remarks
OMP-1	C	1.37 ±0.02	2.69 ±0.02	1.96	Contains TBT, TPT; 29.4% Sn by weight.
	U	1.47 ±0.03	2.82 ±0.03	1.92	
OMP-2	C	1.35 ±0.02	2.70 ±0.02	2.00	Contains TBT; 24.8% Sn by weight.
	U	1.42 ±0.01	2.84 ±0.01	2.00	
OMP-4	C	1.38 ±0.02	2.69 ±0.02	1.95	Contains TBT; 30.0% Sn by weight.
	U	1.40 ±0.02	2.92 ±0.02	2.09	
Paint 1020A	C	1.41 ±0.02	3.76 ±0.02	2.67	$(C_4H_9)_3SnF$, 18.1% by weight.
	U	1.424±0.004	3.761±0.008	2.64	$[(C_4H_9)_3Sn]_2O$, 4.2% by weight.
Paint Glidden	C	1.47 ±0.02	3.45 ±0.02	2.35	$[(C_4H_9)_3Sn]_2O$ 9.1% by weight.
	U	1.447±0.007	3.35 ±0.01	2.32	
Alum-A-Tox	U	1.41 ±0.01	3.56 ±0.02	2.52	$(C_4H_9)_3SnF$, 7.1% by weight.

*Abbreviations: OMP = organotin polymer, TBT = tri-n-butyltin, TPT = tri-n-propyltin, C = cured, U = uncured.

gen. A liquid cell consists of a 3mm thick, 38mm x 51mm aluminum plate with a 13mm hole in the middle for holding the sample. Mylar windows are glued onto the plate and the solution is placed into the cell through a slot. The frozen solution state is assumed to represent an average description of the solution state at any one instant of time. All liquid samples were quick frozen in liquid nitrogen; crystallization is likely for pure liquids.

The organotin polymers were donated by E. Fischer of the D. Taylor Naval Ship Research and Development Center. OMP-1 is poly(tri-n-butyltin methacrylate/tri-n-propyltin methacrylate/methylmethacrylate), OMP-2 is poly(tri-n-butyltin methacrylate/methylmethacrylate), and OMP-4 is the tri-n-butyltin ester of poly(methylvinylether/maleic acid).

The coating designated "Glidden" is a conventional vinyl coating which contains $[(C_4H_9)_3Sn]_2O$ as toxicant. The 1020A coating contains vinyl resin, $(C_4H_9)_3SnF$, and $[(C_4H_9)_3Sn]_2O$; it is the Mare Island Paint Laboratory Formula 1020A. The Alum-A-Tox coating is also a conventional vinyl coating which contains $(C_4H_9)_3SnF$ as toxicant. Since the coating matrix was also part of the study, several coatings were prepared using any one of the above coatings but which did not contain any toxicant. Toxicant was added to the coating so that tin would be 3.5% by weight. Mössbauer measurements were made on cured and uncured coatings. The coatings were cured by depositing thick films on Plexiglas ® and letting the samples air-dry for at least four weeks.

III. RESULTS AND DISCUSSION

It has been suggested in the literature that at low temperatures the organotin halides and hydroxides have structures similar to $(CH_3)_3SnF$. This compound is polymeric with trigonal bipyramidal ligand geometry about the tin atom (5). In Table I one can see that most of the ρ values are above 2.1 in agreement with this suggestion. Even though Ph3SnCl has been found to have a tetrahedral structure at room temperature, it is also expected to form polymeric coordination chains (pentacoordinate tin) at low temperatures in spite of its low ρ value (6). One can argue that as the R group becomes more bulky, steric hindrance will prevent association; the Mössbauer data indicate that association is always present but to a variable degree, depending on the R group.

The crystal structure of $(CH_3)_3SnOAc$ indicates that polymeric chains exist and that the tin is pentacoordinate (7). The ρ values for Ph3SnOAc, $(C_4H_9)_3SnOAc$, $(C_4H_9)_3SnOAcr$, and $(C_4H_9)_3SnOMacr$ in Table I would indicate that these compounds also exist in the same pentacoordinate form although one might have expected significant steric hindrance from the bulky phenyl and butyl groups. Apparently the steric factors are rendered insignificant because of the large Sn-Sn distances in the polymeric chains.

Infrared studies have shown $[(C_3H_7)_3Sn]_2O$, $[(C_4H_9)_3Sn]O$, and $[(Ph)_3Sn]_2O$ to have four-fold coordination about each tin atom (8). The Mössbauer measurements support this assertion as can be seen from the low ρ values for these compounds in Table I.

In interpreting the Mössbauer data on the coatings of interest, it must be kept in mind that the measurements are made at 77^OK and not at room temperature. Although structures can be postulated for an organotin species at a low temperature, the extrapolation to room temperature may not be possible. As an example, Ph_3SnCl is tetrahedral at room temperature but has been postulated to be pentacoordinate at low temperatures (6). Mössbauer techniques should be limited to identifying components and to specifying structures only under the conditions where no phase transitions are expected.

Upon comparing the Mössbauer parameters of the tin species in the organometallic polymers in Table II to the organotin compounds in Table I, it is noted that none of the $(C_3H_7)_3SnX$ or $(C_4H_9)_3SnX$ compounds have Mössbauer parameters similar to those of the $(C_4H_9)_3Sn-$ or $(C_3H_7)_3Sn-$ moieties in the organotin polymers. The ρ values are inconclusive as to the coordination of the tin. A reduction in the magnitude of the ρ values for the organometallic polymers versus the pure $(C_4H_9)_3SnOAcr$ and $(C_4H_9)_3SnOMacr$ compounds is to be expected since the pendant organotin groups are postulated to be far apart in the polymers. The coordination about the tin atom is expected to be four and the ρ value is expected to be low. A reduction in the ρ value is seen; however, it is higher than one would expect for simple four-fold coordination.

The results of measurements on the commercial 1020A coating match those of $(C_4H_9)_3SnF$, indicating no change in this organotin compound upon incorporation into the coating. The peaks for $[(C_4H_9)_3Sn]_2O$ are not observed in the 1020A coating. As shown below the $[(C_4H_9)_3Sn]_2O$ may be chemically transformed into a species having Mössbauer parameters similar to those of $(C_4H_9)_3SnF$ thus obscuring the presence of this starting material. The spectrum of the commercial Glidden coating which contains approximately 10% $[(C_4H_9)_3Sn]_2O$ is interesting in that the parameters are significantly different from the parent compound (see Figure 2), indicating extensive chemical modification of the organotin species in the coating. Since the Mössbauer parameters of $(C_4H_9)_3SnF$ are similar to those of $(C_4H_9)_3SnF$ in the Alum-A-Tox coating, the compound is not altered when it is mixed into the coating.

The Mössbauer parameters for pure $(C_4H_9)_3SnF$ and for this compound mixed into the 1020A, Glidden, and Alum-A-Tox vinyl coatings are given in Table III. Both the isomer shifts and the quadrupole splittings of $(C_4H_9)_3SnF$ in the coatings are very similar to those of the pure compound. This result suggests that small crystals of $(C_4H_9)_3SnF$ remain intact in all

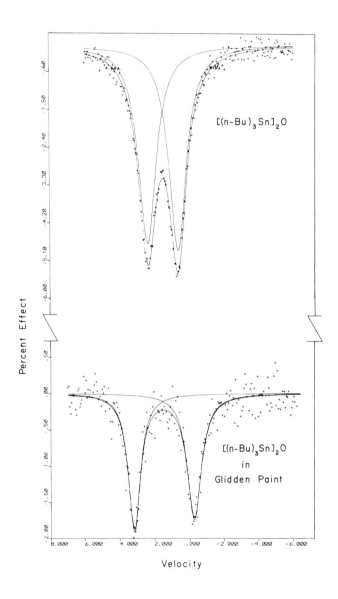

Figure 2. Mössbauer spectra of $[(C_4H_9)_3Sn]_2O$ and $[(C_4H_9)_3Sn]_2O$ in the commercial Glidden coating.

of the matrices. When $(C_4H_9)_3SnF$ is mixed into the 1020A coating containing no pigment or toxicant, a suspension is formed which does not appear to be homogeneous. The suspended material rapidly settles.

TABLE III

$(C_4H_9)_3SnF$ in Various Matrices

Matrix	Isomer Shift (mm/sec)	Quadrupole Splitting (mm/sec)	Remarks
$(C_4H_9)_3SnF$	1.422+0.002	3.871+0.006	
1020A	1.417+0.005	3.77 +0.01	9.4% $(C_4H_9)_3SnF$
Glidden	1.416+0.009	3.72 +0.02	9.4% $(C_4H_9)_3SnF$
Alum-A-Tox	1.42 +0.02	3.68 +0.03	9.4% $(C_4H_9)_3SnF$

In Table IV the Mössbauer parameters for $(C_4H_9)_3SnOAc$ in various matrices are presented. All of the isomer shifts and quadrupole splittings are very similar, regardless of the matrix. Okanara and Ohara (9,10) have shown that triethyl- and tripropyltin formates are associated when dissolved in cyclohexane. Simons and Graham have also shown that trimethyltin formate is associated in methylene dibromide but that the acetate is monomeric in dilute solutions (11). Since the matrix apparently has no affect on the Mössbauer parameters, one can conclude that $(C_4H_9)_3SnOAc$ is associated in the solid state, in the frozen solutions studied, and in the coatings. Upon mixing $(C_4H_9)_3SnOAc$ into the 1020A coating containing no pigment or toxicant, a translucent solution is obtained. The toxicant is widely dispersed in the matrix and the mixture appears to be homogeneous. In the solid state $(C_4H_9)_3SnOAc$ would therefore be a high polymer and in solutions it would a low polymer.

As can be seen from Table V the quadrupole splitting of the organotin species in the 1020A, Glidden, or in the Alum-A-Tox coating to which $[(C_4H_9)_3Sn]_2O$ has been added does not match the quadrupole splitting of pure $[(C_4H_9)_3Sn]_2O$. This suggests a structural change of $[(C_4H_9)_3Sn]_2O$ upon incorporation into these coatings. Considering the magnitude of the quadrupole splitting of the organotin species in the 1020A coating, it is not surprising that the presence of this compound in the commercial 1020A coating was obscured by the $(C_4H_9)_3SnF$ which is present in greater quantity.

TABLE IV

$(C_4H_9)_3SnOAc$ in Various Matrices

Matrix	Isomer Shift (mm/sec)	Quadrupole Splitting (mm/sec)	Remarks
$(C_4H_9)_3SnOAc$	1.439 ± 0.003	3.531 ± 0.006	
1020A	1.439 ± 0.006	3.54 ± 0.01	10.6% $(C_4H_9)_3SnOAc$
Glidden	1.44 ± 0.01	3.49 ± 0.02	10.6% $(C_4H_9)_3SnOAc$
Alum-A-Tox	1.453 ± 0.008	3.55 ± 0.02	10.6% $(C_4H_9)_3SnOAc$
CHCl$_3$	1.45 ± 0.01	3.63 ± 0.03	10% $(C_4H_9)_3SnOAc$ Frozen Solution
n-butylacetate	1.464 ± 0.007	3.54 ± 0.01	9.97% $(C_4H_9)_3SnOAc$ Frozen Solution

A comparison of the Mössbauer parameters of pure $[(C_4H_9)_3Sn]_2O$ to this compound in various frozen solutions (see Table V) indicates that a considerable solvent inter- action is present. The $[(C_4H_9)_3Sn]_2O$ species in the Alum-A- Tox coating is probably the same solvated species as is present in the xylene-methyl isobutyl ketone solution which is the solvent for that coating, but it is not the same as the organotin species in the other coatings or frozen solutions. Infrared studies indicate that an organotin carboxylate is formed in the 1020A and Glidden coatings (12). The Mössbauer studies have neither confirmed nor refuted this finding. The unusual behavior of $[(C_4H_9)_3Sn]_2O$ in these coatings and sol- vents is the object of current research.

These initial results indicate the potential of Mössbauer spectroscopy for identifying organotin components of various marine antifouling coatings and for determining the structure of the organotin species in these matrices. Further delinea- tion of the coatings and model chemical systems is presently underway as is an evaluation of the Mössbauer backscatter technique for determining the organotin species on the surface as opposed to those found in the bulk samples.

TABLE V

$[(C_4H_9)_3Sn]_2O$ in Various Matrices

Matrix	Isomer Shift (mm/sec)	Quadrupole Splitting (mm/sec)	Remarks
$[(C_4H_9)_3Sn]_2O$	1.256+0.003	1.490+0.008	
1020A	1.439+0.007	3.38 +0.01	10.8% $[(C_4H_9)_3Sn]_2O$
Glidden	1.462+0.007	3.28 +0.01	11.2% $[(C_4H_9)_3Sn]_2O$
Alum-A-Tox	1.362+0.006	3.06 +0.01	11.0% $[(C_4H_9)_3Sn]_2O$
m-xylene	1.39 +0.02	2.67 +0.04	9.3% $[(C_4H_9)_3Sn]_2O$ Frozen Solution
Xylenes Methyl-isobutylketone	1.29 +0.01	2.91 +0.03	10.9% $[(C_4H_9)_3Sn]_2O$ Frozen Solution
Benzene	1.32 +0.01	2.55 +0.02	10.2% $[(C_4H_9)_3Sn]_2O$ Frozen Solution

IV. ACKNOWLEDGEMENT

This work is a result of research sponsored by the Office of Naval Research and the NOAA Office of Sea Grant, Department of Commerce, under Grant #R/MTR-1. Several organotin compounds were generously donated by M&T Chemicals, Inc. The Glidden coatings, with and without toxicant, were donated by Glidden-Durkee. The Alum-A-Tox coatings, with and without toxicant, were contributed by Standard Paint and Varnish Company of New Orleans. The 1020A coatings with toxicant, without toxicant, and without toxicant and pigment were donated by the U. S. Navy, and the organometallic polymers were contributed by the D. Taylor Naval Ship Research and Development Center.

V. REFERENCES

1. Greenwood, N. N., and Gibb, T. C., Mössbauer Spectroscopy," Chapman and Hall, Ltd., London, 1971, pp. 46-59.
2. Herber, R. H., Stockler, H. A., and Reichle, W. T., J. Chem. Phys., 42, 2447 (1965).
3. Reichle, W. T., Inorg. Chem., 5, 87 (1966).
4. Zuckerman, J. J., Adv. Organometal. Chem., 9, 21 (1970).

5. Clark, H. C., O'Brien, R. J., and Trotter, J., J. Chem. Soc., 2332 (1964).
6. Bokii, N. G., Zakharova, G. N., and Struchkov, Yu. T., J. Structural Chem., 11, 828 (1970).
7. Chih, H. and Penfold, B. R., J. Cryst. Mol. Struct., 3, 285 (1973).
8. Kriegsmann, V. H., Hoffmann, H., and Geissler, H., Z. Anorg. Allgem. Chem., 341, 24 (1965).
9. Okawara, R., and Ohara, M., Bull. Chem. Soc. Japan, 36, 624 (1963).
10. Okawara, R., and Ohara, M., J. Organometal. Chem., 1, 360 (1964).
11. Simons, P. B., and Graham, W. A. G., J. Organometal. Chem., 10, 457 (1967).
12. Hoffman, J. F., Kappel, K. C., Frenzel, L. M., and Good, M. L., see preceding paper.

PRODUCTION OF BIOMEDICAL POLYMERS

I. SILICONE/URETHANE SYNERGY

IN AVCOTHANE[R] ELASTOMERS

Robert S. Ward, Jr.
Avco Medical Products

Emery Nyilas
Avco Everett Research Lab

ABSTRACT

The combination of a reactive poly(dimethylsiloxane) with a segmented poly(etherurethane) gives improved blood compatibility as well as enhanced physical/mechanical properties to Avcothane 51 elastomer. After the clinical application of 30,000 intra-aortic balloon pumps fabricated from this silicone/urethane hybrid, no clinically intolerable hematologic effects have been demonstrated. Performance superior to either homopolymer is attributed to an increased concentration of dispersion type bonding groups in the polyurethane surface at the expense of high-energy polar binding sites. This results in favorable, low energy interactions with plasma proteins, the first species to populate any surface exposed to blood. At 10 wt. % concentration, the silicone does not inhibit interchain attraction in the urethane; tensile strengths are actually greater than those for the urethane alone. SEM and dynamic mechanical testing indicate a molecular weight-dependent, two-phase morphology typical of block copolymers and polyblends. Lack of phase separation in Avcothane 51 prepolymer solutions is apparently related to a surfactant effect by a block copolymer, stabilizing unreacted silicone micelles. The effects of drying conditions on the surface composition of solvent-cast films have been shown, in turn, to give rise to varying degrees of apparent hemocompatibility. Optimization requires control of the degree of copolymerization as measured by GPC, and characterization of surfaces with respect to the proportions of silicone and urethane present. Using IRATR, SEM and electron microprobe analysis, an "IRATR Index" has been defined and shown by independent ex vivo and in vitro techniques to be correlatable with overall hemocompatibility. Thus, through strict production controls, cardiac assist devices having high quality blood-contact surfaces are routinely prepared, which do not rely on incorporated anticoagulants and exhibit the required engineering properties.

I. INTRODUCTION

Synthetic polymers are the materials of choice for a wide variety of implantable biomedical devices. Elastomers, in particular, possess the mechanical properties required of blood pumps, catheters, and related cardiovascular products. Yet, nearly all commercially available resins induce a substantial degree of hematologic damage when used in contact with blood.

Modern surgical techniques, such as cardiopulmonary bypass, cardiac assistance and all types of vascular catheterizations are subject to the adverse effects induced by blood/foreign surface interactions. The safety of such procedures, especially those requiring long-term applications, increases in proportion with the blood compatibility of the contact surfaces.

In addition to being hemocompatible, implantable materials must possess the appropriate engineering properties for each of the particular applications. In the case of blood pumps of the balloon or diaphragm type, this implies good flex life and hydrolytic stability as well. Ease of fabrication and adaptability to volume production methods are additional constraints which determine the applicability of candidate materials.

The mechanism involved in clotting or thrombosis, as induced by surfaces foreign to blood, is poorly understood at present. Consequently, the development of blood-compatible polymers has been approached from many directions. The use of negative surface charges, ionically or covalently bonded heparin, hydrogel coatings and glow discharge treatments have been reported to reduce thrombogenicity. While a recent review (1) has surveyed the current schools of thought on foreign surface-induced thrombosis, the fact remains that very few polymers are currently available in quantities sufficient to support production of implantable prosthetic devices.

Despite differences in approach, it is generally agreed that, when blood contacts a foreign surface, initial events include the adsorption of plasma proteins (2, 3). This is the result of the higher diffusivity and bulk concentration of the plasma proteins with respect to the cellular elements, and has been confirmed experimentally (4, 5). Since subsequently arriving platelets and white blood cells encounter not the original surface but a mat of adsorbed protein, it has been assumed that the conformational state of this layer is particularly critical in the prevention of thromboembolic phenomena.

In some cases, Baier has achieved good correlation (6) between blood compatibility and the critical surface tension, γc, as determined from Zisman plots (7). The optimum lies within a specific range of low (viz., 20-30 dyne/cm) γc values, which may correspond to 'gentle' interactions with plasma proteins. It follows that, if during adsorption, native plasma proteins become affected by surface groups capable only of low energy interactions, then the structural alterations induced may be insufficient to induce thrombosis. The result will be a dynamic passivation of the polymer surface by the sorbed protein layer.

To encourage this passivation, surfaces can be employed which are only capable of exerting low energy van der Waals/London type forces. However, these surface properties must be accomplished within the other constraints placed on the polymer, in order to produce a practically useful biomaterial.

II. POLY(DIMETHYLSILOXANES)

Pure poly(dimethylsiloxanes) (DMS) are known to produce films with low γc values associated with a cloud of methyl groups oriented toward the air-facing surface (8). An unusually low level of intermolecular forces also results in poor mechanical properties in the absence of reinforcing fillers. Fumed silica, the most effective filler for silicone, is highly thrombogenic due to its polar silanol groups acting as potential H-bonding sites for plasma proteins. The resulting filled polymer can also be variably thrombogenic depending upon the amount of silica exposed in the surface (9).

When combined with organic polymers as a minor constituent, the silicones are naturally surface active (8). They are generally incompatible with the organic phase, tending to be excluded from the bulk, and are able to wet non-halogenated organic resins, making spreading a spontaneous process.

III. POLY(ETHER URETHANE)

The segmented polyurethanes are the reaction products of polymeric ether glycols and a slight excess of a diisocyanate, and a "chain extending" diamine or dihydroxy derivative. They are block copolymers composed of "soft" blocks (i.e., low Tg) of polyether, and short "hard" segments of urethane, with or without urea functionality, giving rise to a high degree of intermolecular hydrogen bonding. By employing various types of isocyanates (aliphatic, aromatic), glycols (polyethylene, polypropylene, polytetramethylene) and chain extenders (aliphatic, aromatic), a wide variety of structures is possible. The common feature of these urethanes is a fine, two-phase micromorphology arising from the thermodynamic incompatibility of the soft and hard segments.

This structure can give rise to thermolabile or 'virtual' crosslinking since the hard urethane domains act as tie points at temperatures below their crystalline melting points but allow the polymer to flow at higher temperatures. The result is a polymer which is a 'snappy' elastomer, effectively vulcanized at use temperatures, but which is thermoplastic at elevated temperatures. If stoichiometry during the condensation type polymerization is carefully balanced, the product may be solvent-soluble as well.

Some 'segmented' polyurethanes possess physical properties and in vivo stability well suited to the fabrication of cardiovascular devices, particularly the more demanding applications such as flexing blood pumps. Typical compositions give γc in the range of 28-35 dyne/cm, higher than the 22 dyne/cm of methylsilicones and indicative of somewhat higher surface energy.

IV. AVCOTHANE 51 ELASTOMER

Avcothane 51 elastomer is a hybrid of a poly(ether urethane) (90%) and poly(dimethylsiloxane) (10%), which retains the excellent engineering properties of the urethane while displaying a silicone-like surface. In more than 30,000 clinically implanted Avco Intra Aortic Balloon Pumps, this elastomer exhibited excellent blood compatibility which has been verified by statistical analyses published elsewhere (11). Sixteen-week immersion in Ringers solution at 45°C did not produce any significant change in tensile strength, ultimate elongation, or modulus. The well-known hydrolytic stability of poly(ether urethane) is thought to be enhanced by the presence of the hydrophobic silicone.

TABLE 1

COMPARATIVE PROPERTIES OF AVCOTHANE   51

ELASTOMER AND ITS POLYETHERURETHANE

	AVCOTHANE^R 51	PEU
Density (g/cm^3)	1.09	1.11
Tensile Strength (psi)	6200	5700
Ultimate Elongation (%)	580	520
Indentation Hardness (shore "A")	72	80
Graves Tear Resistance (lb/in.)	490	400
Dielectric Strength (volts/mil)	1500	550

A. Preparation

If a common, methyl-terminated DMS is physically combined with an anhydrous solution of a segmented urethane, phase separation will quickly result, as suggested by the large difference in the solubility parameters of the two polymers (Δ=2.7). Using a high shear mixer, a temporary emulsion can be formed. Upon casting a film from this emulsion, the silicone micelles coalesce and bleed from the urethane film following solvent evaporation. In addition, the urethane's tensile strength is severely degraded at silicone concentrations above the fractional weight percent level.

By using diacetoxy-terminated siloxane instead of the silicone oil, some branched block copolymer of urethane and silicone can be formed by condensation reaction between the terminal hydroxy groups of the urethane and the silicone's acetoxy functionalities. Acting as a surfactant, this copolymer appears to stabilize the remaining unreacted silicone/urethane mixture. Solutions of Avcothane 51 elastomer remain 'homogeneous', though opaque, after months of storage in sealed containers.

B. Morphology

When an Avcothane 51 film is cast, atmospheric moisture initiates the self-catalyzed curing reaction as in a one-component silicone RTV system. A lightly crosslinked composite results, in which comparatively large vulcanized silicone domains are dispersed in a urethane-rich continuous phase, with some covalent linkages between the two. The spherical silicone domains have a disperse size distribution which is dependent upon the molecular weight of the unvul-canized silicone, as shown in Fig. 1. Even with a siloxane

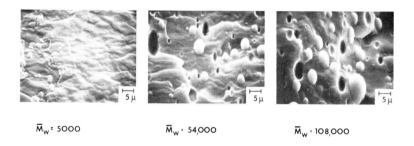

$\overline{M}_W = 5000$ $\overline{M}_W = 54,000$ $\overline{M}_W = 108,000$

Fig. 1. Two Phase Morphology of Avcothane 51 Elastomer as a Function of Silicone Molecular Weight.

having a molecular weight of only 5,000, domain size is large enough to scatter light and make the films translucent.

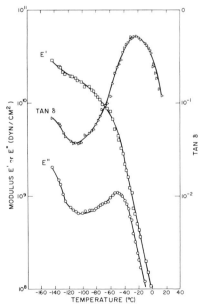

Figure 2 gives the results of dynamic mechanical test-ing obtained on Rheovibron at 110 Hz and 1°C/min heating rate. The sample tested is equivalent to the '54,000' specimen shown in Fig. 2. Two glass transitions are apparent, corresponding roughly to those of the sili-cone (-130°C) and the ure-thane soft segment (-50°C), which is consistent with the block copolymer or polyblend structure of Avcothane 51 elastomer. The lack of a well-defined plateau between the two Tg's may indicate some phase mixing, despite the obvious silicone domains revealed by the SEM in Fig. 1.

Fig. 2. Dynamic Mechanical Test (Rheovibron) of Avcothane 51 Elastomer; 110 Hz, 1°C/min.

Although properly prepared surfaces of Avcothane 51 elastomer show no such silicone domains, their critical surface tensions are identical to pure silicone. Low molecular weight silicone fractions have greater compatibility with the urethane and hence, they can exist within the urethane-rich continuous phase.

C. Tensile Strength

The ultimate tensile strength of Avcothane 51 is ~6200 psi. Assuming 100 psi tensile strength for the vulcanized unfilled silicone homopolymer and 5700 psi for the polyurethane, the weighted average for their 90/10 mixture is 5140 psi, or 17% less than actual. The reason for the significant increase in tensile strength has not been determined. The silicone certainly has not reduced interchain hydrogen bonding in the urethane, although other less flexible covalent crosslinks are known to interfere with these virtual crosslinks and can actually reduce tensile strength, unless present at high densities (12).

D. Surface Properties

The polyblend nature of Avcothane 51 prepolymer solutions requires careful control to achieve optimum surface composition during the fabrication of blood-contacting devices. The coupling of the polyfunctional DMS with the essentially difunctional, hydroxy-terminated polyurethane will result in the formation of a gel phase if allowed to proceed to high extents of reaction. Insoluble gel present in the solution will adversely affect surface morphology and composition of the cast film.

Gel permeation chromatography (GPC) is used to follow the copolymerization reaction. Formation of a strongly UV-absorbing, high MW shoulder on the original polyurethane molecular weight distribution is taken as evidence of copolymerization since branched DMS would be nearly transparent to UV at the detector sensitivities employed. An optimum degree of copolymerization has been found to exist; enough to stabilize the emulsion but not enough to degrade surface properties of the film. This relationship is shown in Fig. 3.

In addition to the degree of copolymerization, several other variables are known to affect the surface properties of Avcothane 51 elastomer films solvent-cast under ambient conditions. The surface active properties of the silicone component and its incompatibility with the urethane make the concentration of silicone anisotropic within the cured film.

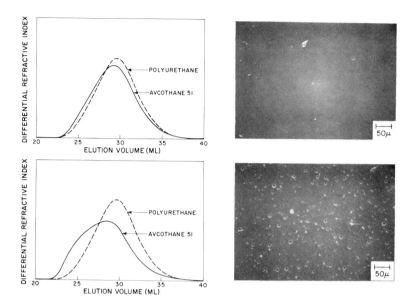

Fig. 3. Dependence of Surface Morphology of Cast Films on Extent of Reaction in Prepolymer Solution.

Rate of solvent evaporation, the nature of the substrate, and atmospheric humidity all affect this distribution. Since a particular ratio of silicone-to-urethane has been found to yield optimum hemocompatibility (3), these variables have a direct effect on the in vivo performance of Avcothane 51 blood contact surfaces.

Infrared attenuated total reflection (ATR) spectroscopy is used to measure the ratio of the two polymer components present in the blood-contact surface, and is routinely employed as a quality control technique. An "IRATR Index" has been defined (Fig. 4), and found to be inversely proportional to the concentration of silicone in the surface, as determined by electron microprobe analysis. Figure 5 shows this correlation.

V. SUMMARY

The low energy surface required for blood compatibility and the strong intermolecular forces resulting in good 'physicals' are achieved in a soluble elastomer by incorporating a reactive   DMS into a poly(ether urethane). The two-phase morphology of the system appears to satisfy what,

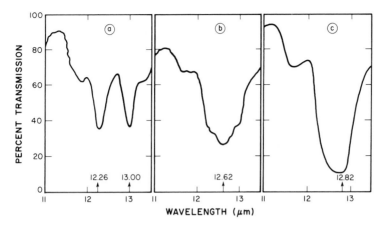

Fig. 4.   a.   Polyetherurethane   b.   Polydimethylsiloxane
c.   Composite IRATR Spectrum.   IRATR Index = A13.0μm +
A12.62μm.

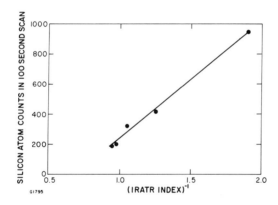

Fig. 5.   Correlation between IRATR INDEX and ELECTRON
MICROPROBE as measurement techniques for silicon(e)
surface concentration. r^2=0.97.

for a single homopolymer, might be mutually exclusive
requirements.   The sensitivity of surface topology and composition to the variables mentioned, is also a result of this
morphology, the stability of which is reduced by the large
solubility parameter difference between the two different
polymers.   While necessitating controlled conditions during
device fabrication, this basic incompatibility of silicone with
organic resins is largely responsible for the unique combination of properties of Avcothane 51 elastomer.
   The Avco Intra Aortic Balloon Pump has displayed a high

degree of blood compatibility in thousands of human implan-
tations. However, any potentially hemocompatible polymer
may be made thrombogenic through contamination and poor
handling. Ex vivo experiments have shown (14) that even
micron-sized dust particles or other surface inhomogen-
eities can produce 'wedge' thrombi when present in blood
contact surfaces. Clean room technique and special methods
of material handling are required to optimize performance
and minimize hematologic trauma to the patient. Thus, the
favorable in vivo performance of Avco Intra Aortic Balloon
Pumps is as much a result of careful manufacturing procedures as it is of the properties of Avcothane 51 elastomer
itself.

VI. REFERENCES

1. Vroman, L., and Leonard, E.F., editors, "The Be-
havior of Blood and Its Components At Interfaces", Ann.
N.Y. Acad. Sci., 283, 1977.
2. Baier, R.E., "The Organization of Blood Components
Near Interfaces", Ann. N.Y. Acad. Sci., 283, 17, 1977.
3. Nyilas, E., and Ward, R.S., "Development of Blood-
Compatible Elastomers V. Surface Structure and Blood
Compatibility of Avcothane Elastomers", J. Biomed.
Mater. Res. Symposium, 8, 69, 1977.
4. Dutton, R.C., Baier, R.E., Dedrick, R.L., and
Bowman, R.L., Trans. Amer. Soc. Artif. Int. Organs,
14, 57, 1968.
5. Petschek, H.E., Adamis, D., and Kantrowitz, A.R.,
Trans. Amer. Soc. Artif. Int. Organs, 14, 256, 1968.
6. Baier, R.E., "Surface Properties Influencing Biological
Adhesion", in "Adhesion in Biological Systems", Ann.
N.Y. Acad. Sci., 15, 1970.
7. Zisman, W.A., "Relation of the Equilibrium Contact
Angle to Liquid and Solid Constitution", in "Contact
Angle, Wettability and Adhesion", Fowkes, F.M., and
Gould, R.F., editors, 1, Advances in Chem. Series 43,
Amer. Chem. Soc., Washington, DC, 1964.
8. Noll, W., "Chemistry and Technology of Silicones", 324,
Academic Press, New York, 1968.
9. Nyilas, E., Burnett, P., Haag, R.M., and Kupski,
E.L., "Surface Microstructural Factors and the Blood
Compatibility of a Silicone Rubber", J. Biomed. Mater.
Res., 4, 368, 1970.
10. Boretos, J.W., and Pierce, W.S., "Segmented Poly-
urethane: A New Elastomer for Biomedical Applica-
tions", Science, 158, 1481, 1967.
11. Nyilas, E., Leinbach, R.C., Caulfield, J.B., Buckley,
M.J., and Austin, W.G., "Development of Blood Com-

patible Elastomers. III. Hematologic Effects of Avcothane Intra-Aortic Balloon Pumps in Cardiac Patients",
J. Biomed. Mater. Res. Symposium, 3, 129, 1972.
12. Saunders, J.H., and Frisch, F.C., "Polyurethanes:
Chemistry and Technology", Vol. No. I., p. 295, Wiley
Interscience, New York, 1962.
13. Nyilas, E., and Ward, R.S., Polymer Preprints, 16,
#2, 165, 1975.
14. Morton, W.A., and Cumming, R.D., "A Technique for
for the Elucidation of Virchow's Triad.", Ann. N.Y.
Acad. Sci., 283, 477, 1977.

NEW POLYORGANOSILOXANES PREPARED FROM
INORGANIC MINERAL SILICATES

R.Atwal, B.R.Currell, C.B. Cook,
H.G.Midgley and J.R. Parsonage

Thames Polytechnic, London, U.K.

Polyorganosiloxanes may be prepared by the trimethyl-silylation of mineral silicates. The structure of the mineral, in particular the degree of isomorphous replacement of Si by Al, controls the extent of reaction and also the molecular weight of the product. By heat treatment the molecular weight of the poly-organosiloxanes may be increased and products containing hydroxyl groups obtained. Further reactions of the hydroxyl groups have been investigated as a basis for the preparation of new high molecular weight materials.

I. INTRODUCTION

The technique of trimethylsilylation developed by Lentz (1) involves the reaction of silicate materials with a mixture of concentrated hydrochloric acid, water, isopropanol and hexamethyldisiloxane. This mixture converts the silicate anions into trimethylsilyl derivatives whose molecular weight can be determined; they can also be separated and analysed by conventional chromatographic techniques.

$$^-O.\!\!-\!\!\underset{\underset{O^-}{|}}{\overset{\overset{O^-}{|}}{Si}}\!\!-\!\!O\!\!-\!\!\underset{\underset{O^-}{|}}{\overset{\overset{O^-}{|}}{Si}}\!\!-\!\!O\!\!-\!\!\underset{\underset{O^-}{|}}{\overset{\overset{O^-}{|}}{Si}}\!\!-\!\!O^-$$

$$\downarrow$$

$$XO\!\!-\!\!\underset{\underset{OX}{|}}{\overset{\overset{OX}{|}}{Si}}\!\!-\!\!O\!\!-\!\!\underset{\underset{OX}{|}}{\overset{\overset{OX}{|}}{Si}}\!\!-\!\!O\!\!-\!\!\underset{\underset{OX}{|}}{\overset{\overset{OX}{|}}{Si}}\!\!-\!\!OX$$

$$X = SiMe_3$$

This general method has been applied to an attempted
measurement of silicate chain lengths in sodium silicate
solutions and portland cement pastes by Lentz (1), chain
lengths in silicate glasses by Masson (2), and the
structure of various sodium silicate hydrates by Glasser (3).

The scope of the reaction was extended by Kenney et al
(4,5) who have prepared organosilicon polymers derived from
chrysotile and apophyllite; in the former case the polymers
had a fibrous nature similar in size, and diameter length
ratio to the parent chrysotile. Fripiat and co-workers (6)
have also been active in the preparation of organosilicate
materials.

In the United Kingdom, the Paint Research Association (7),
and also Currell et al (8,9) have applied the trimethylsilylation technique to a range of minerals in an attempt
to prepare new soluble polyorganosiloxanes whose macromolecular character mirrors that of the starting mineral
silicate. Materials similar to the conventional silicones
have been prepared and the PRA have shown that these have
considerable potential in the waterproofing of masonry and
textiles, in the modification of alkyl resins, and as
additives to prevent pigment separation in tinted paints (7).

II. RESULTS AND DISCUSSION
A. The Relationship Between Mineral Structure And
Polyorganosiloxane Produced.
     We have previously reported the application of the Lentz
trimethylsilylation technique to a wide range of inorganic
mineral silicates (8,9). Many mineral silicates undergo no
reaction whatsoever on heating with the mixture of hexamethyldisiloxane, isopropanol, water and hydrochloric acid
at reflux temperatures, whilst others undergo complete
dissolution and reaction. The object of the work reported
here was to delineate the factors controlling the reactivity
of minerals in the reaction media, and also to try and
correlate the structure of the mineral used with the type of
polyorganosiloxane produced.

Mineral silicates with two distinct characteristics
appear to react under these conditions, firstly the inter-
layer cations must be easily leachable by hydrochloric acid
and secondly, a high percentage of aluminium must be present
in the silicate backbone. Therefore biotite, phlogopite and
muscovite have been found to be particularly suitable for
the production of polyorganosiloxanes. Cation removal is
aided by the presence of heavy metal ions such as iron, and

a loosely packed silicate backbone. Thus in single chain
silicate minerals diopside undergoes no reaction whereas
wollastonite was found to be reactive. Vermiculite with
relatively little isomorphous replacement of silicon by
aluminium in the silicate backbone underwent reaction to
give insoluble material of high molecular weight (4000),
whereas the micas mentioned above have Al:Si::1:3 and give
soluble products from the trimethylsilylation reaction. We
would postulate that acid attack on the silicate backbone
occurs at the aluminium centres thus breaking the sheet at
these points to allow attack by trimethylsilanol. This
gives a resultant polymer ($M_n \sim 900$) soluble in organic
solvents. Further experience has confirmed the importance
of aluminium in these reactions; in the trimethylsilylation
of a series of calcium aluminosilicate glasses the yield of
polymer produced decreased with decreasing aluminium content
(see Table 1). Even when the percentage silicon in the
glass is very low (SiO_2 6.1%) the presence of 50% Al_2O_3
enabled us to react 10% by weight of the glass.

TABLE 1

The Trimethylsilylation Of Some
Calcium Aluminosilicate Glasses

Glass Composition			Polymer yield[*]
SiO_2	Al_2O_3	CaO	(%)
36.9	26.8	36.3	89.1
64.0	9.1	26.0	1.5
90.0	5.0	5.0	0.9

*Calculated on yield of unreacted glass

   The trimethylsilylation of the mineral phlogopite has
been investigated in detail and the yield of polymer
produced at any particular reaction time is directly
proportional to the weight of aluminium leached from the
mineral. The unreacted mineral, even after extensive
reaction times, is identical in chemical composition and
almost identical in chemical structure with the starting
mineral. There is no preferential leaching of any specific
cation and the reaction appears to occur at the mineral
surface; the unreacted mineral has a significantly larger
average particle size (>3μ) compared with the starting
mineral (<½μ) indicating the reaction depends on particle
size. Phlogopite (Harty Mountain Range, Western Australia)

formula

$$(Ca,Na,K,Rb)_{1.995}(Mg,Fe,Ti,Al,Mn)_{6.056}$$

$$(Si_{5.658}Al_{2.341})O_{20.215}(OH,F)_{3.785}$$

undergoes 80% reaction after 168 hours; after an initial
accelerated reaction, cations are removed at a uniform
rate and the polyorganosiloxane product is of constant
composition and independent of reaction time (see figure 1).

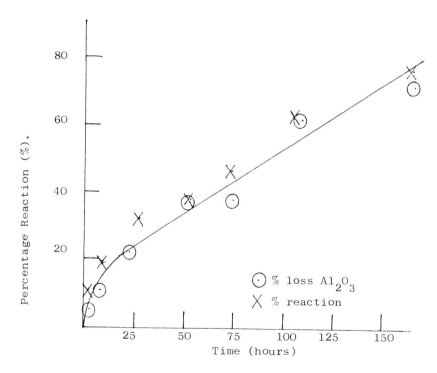

Fig 1: The trimethylsilylation of phlogopite

B.   Structure Of Polyorganosiloxanes And Their Control
By Heat.

The polyorganosiloxanes prepared by the trimethylsilylation of the mineral biotite have been examined in
detail.   The polyorganosiloxane is initially obtained by
physical separation from the aqueous phase, unreacted
mineral and the hexamethyldisiloxane/isopropanol mixture
followed by a first treatment at 60°C/17mmHg and a second
treatment at 45° C/3 mmHg pressure to constant weight to
remove hexamethyldisiloxane and any residual isopropanol.
The resultant polymer (M_n 900) consists of 47% tetrakis-
trimethylsiloxysilane (I) and 6% hexakistrimethylsiloxy-
disiloxane (II); the remainder consists of straight chain
and cyclic homologues.   These higher molecular weight
fragments of varying percentage hydroxyl content are in the
molecular weight range up to 4000.

If the polymer is heated at 100°C for five hours a
weight loss is observed caused by the loss of I (b.p. 228°C)
and II (b.p. 303°C).   This is accompanied by an increase in
percentage of hydroxyl groups.   At higher temperatures the
loss of low molecular weight analogues and increase in
hydroxyl group content is observed until at about 180°C
hydroxyl condensation occurs to give high molecular weight
analogues. The results of these investigations are
illustrated in Table 2 and Figure 2.

TABLE 2

Polymers Prepared From The Mineral Silicate

Treatment Temp($^{\circ}$C)	Press (mmHg)	Time (hrs)	Wt.loss (%)	Carbon (%)	Hydrogen (%)
45	3		–	30.96	8.12
100	760	5	2.5	30.75	8.07
200	760	5	35.3	27.92	7.67
235	760	5	47.0	25.85	6.65
300	760	5	62.6	3.52	1.71[+]
500	760	5	68.0	0.00	0.00[+]
700	760	5	66.0	0.00	0.00[+]

[+]insoluble products.

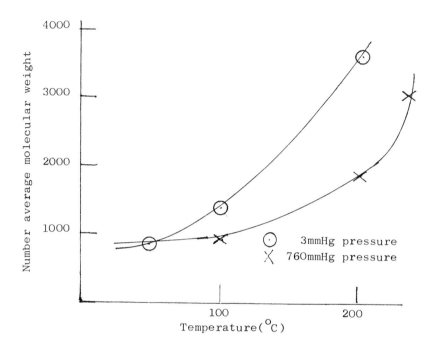

Fig. 2:   The variation of $\overline{M}_n$ with temperature

Above a molecular weight of the order of 4000, the products are insoluble with elemental analysis figures approximating to building units of the type

$$CH_3 - \underset{\underset{CH_3}{|}}{\overset{\overset{CH_3}{|}}{Si}} - O - \underset{\underset{O_{0.5}}{|}}{\overset{\overset{O_{0.5}}{|}}{Si}} - O_{0.5}$$

(Found: C, 25.85; H, 6.65; $C_3H_9SiO_{2.5}$ requires C, 25.53 H, 6.38%)

Products of this type can be found in low yield by the trimethylsilylation of sheet silicates with low aluminium content, e.g. vermiculite.

C. The Reactions Of Hydroxylated Polymers.

The reactivity of the hydroxyl groups attached to the polyorganosiloxane has been investigated with the objective of establishing routes for the preparation of high molecular weight materials, for example the following reactions :

$$\gtrdot Si-OH + Me_3SiCl \longrightarrow \gtrdot Si-O-SiMe_3$$

$$\gtrdot Si-OH + Ph_2PCl \longrightarrow \gtrdot Si-O-PPh_2$$

$$\gtrdot Si-OH + CH_2=CHSiMeCl_2 \longrightarrow \gtrdot Si-O-\underset{\underset{Cl}{|}}{Si}MeCH=CH_2$$

Reactions with acetyl chloride, phenylisocyanate and thionyl chloride have been observed, but complications arise in the redistribution of the silicate backbone. Routes for the preparation of high molecular weight materials may now be developed using the reaction of the hydroxyl groups with difunctional instead of the above monofunctional reagents.

III. SUMMARY

The preparation of polyorganosiloxanes by the trimethyl-silylation of mineral silicates is discussed in terms of the mineral structure. The result of the subsequent heat treatment of these polyorganosiloxanes and the reaction of hydroxyl groups on the siloxane backbone is reported.

IV. ACKNOWLEDGEMENTS

The research reported herein has been sponsored in part by the United States Government. We also acknowledge the gift of calcium aluminosilicate glasses from Corning Glass Works.

V. REFERENCES
1. Lentz, C.W., Inorganic Chemistry. 3, 574 (1964).
2. Gotz, J.W., and Masson, C.R., J.Chem.Soc.(A). 2683 (1970) and ibid 686 (1971).
3. Sharma, S.K., Dent Glasser, L.S., and Masson, C.R., J.Chem.Soc.(Dalton). 1324 (1973).
4. Frazier, S.E., Bedford, J.A., Hower, J., and Kenney, M.E. Inorganic Chemistry. 6, 1693 (1967).
5. Linsky, J.P., Paul, T.R., and Kenney, M.E., J.Polymer Sci. Polymer Phys.Edit., 9, 143 (1971).
6. Zapata, L., Castelein, J., Mercier, J.P., and Fripiat,J.J. Bull.Soc.Chim.Fr., 54 (1972).

7. Paint Research Association. <u>U.K.Patent Appln.No.</u> 40474/73.

8. Currell, B.R., Midgley, H.G., and Seaborne, M.A., <u>J.Chem.Soc.(Dalton)</u>. 490 (1972).

9. Currell, B.R., Midgley, H.G., Seaborne, M.A., and Thakur, C.P., <u>Brit.Polymer J</u>. 6, 229 (1974).

METHYL- AND 3,3,3-TRIFLUOROPROPYL-SUBSTITUTED
m-XYLYLENESILOXANYLENE POLYMERS

Harold Rosenberg and Eui-won Choe*
Air Force Materials Laboratory

The synthesis and characterization of a new class of organosilicon polymers, the poly(m-xylylenesiloxanylenes) with general structure I, was undertaken as part of an investiga-

$R_1 = R_2 = CH_3$ or $CF_3CH_2CH_2$

$R_3 = CF_3CH_2CH_2$ or CH_3

$x = 0, 1$ or 2

tion of new viscoelastic, heat-resistant polymers with potential for development into broad use-temperature, fuel-resistant sealants. Representative members of 3 types of methyl- and 3,3,3-trifluoropropyl-substituted polymers within this class, namely poly(m-xylylenedisiloxanylenes) (I, x = 0), -trisiloxanylenes (I, x = 1), and -tetrasiloxanylenes (I, x = 2), were synthesized from α,α'-bis(hydroxydialkylsilyl)-m-xylenes by homopolymerization, as well as cocondensation reactions with bis-aminosilanes and -disiloxanes. Polymerization products were characterized with regard to chemical composition, molecular weight and glass-transition temperature, and evaluated for thermal stability and hydrocarbon fuel resistance. From the data, structure-property correlations were established to enable prediction of molecular structures of m-xylylene-, and possibly other arylenesiloxanylene polymers with optimum combinations of desired properties.

I. INTRODUCTION

    Although silicones and certain other siloxane-type polymers have been of interest as both broad temperature range as well as chemical-resistant elastomers, little attention has been given to silarylenesiloxane,[1,2] or arylenesiloxanylene,

*Present address: Celanese Research Company, Summit, NJ 07901

polymers for such applications. With the goal of providing a route to new viscoelastic polymers with potential for development into fuel-resistant, wide use-temperature range sealants, an investigation of several classes of arylenesiloxanylene polymers was undertaken. In order to predict molecular structures of poly(arylenesiloxanylenes) which would exhibit the optimum combinations of desired properties, it was found necessary to first develop structure-property correlations. To achieve the latter, we set out to synthesize representative members of several classes of appropriately substituted arylenesiloxanylene polymers and characterize them with respect to their glass-transition temperature, thermal stability and fuel insolubility. From the data thus obtained, it would be expected that desired structure-property correlations could be established. One class of polymers selected for such investigation was that of the poly(m-xylylenesiloxanylenes), I, polymer systems that would be predicted to exhibit elastomeric

$$\left[\begin{array}{c} CH_3 \\ | \\ -SiCH_2 \\ | \\ R_1 \end{array} \bigcirc \begin{array}{c} CH_3 \\ | \\ -CH_2SiO \\ | \\ R_2 \end{array} \left(\begin{array}{c} CH_3 \\ | \\ SiO \\ | \\ R_3 \end{array}\right)_x \right]_n$$

I

$R_1 = R_2 = CH_3$ or $CF_3CH_2CH_2$

$R_3 = CF_3CH_2CH_2$ or CH_3

$x = 0, 1$ or 2

properties over a broad temperature range. In this paper the results of that investigation are briefly reported with regard to the synthesis of the representative methyl- and 3,3,3-trifluoropropyl-substituted polymers, their characterization (in part), and the structure-property correlations which were derived from the data.

II.   RESULTS AND DISCUSSION

A.   Syntheses

In order to obtain appropriately structured and representative members for characterization, three types or families of m-xylylenesiloxanylene polymers were chosen for synthesis. These were the poly(m-xylylenedisiloxanylenes), II; -trisiloxanylenes, III; and -tetrasiloxanylenes, IV. To prepare these polymers, a number of methyl- and 3,3,3-trifluoropropyl-substituted bis-chlorosilanes, bis-hydrosilanes, and bis-silanols, together with bis-aminosilanes and siloxanes, were initially synthesized and characterized.[3] The bis-silanols were homopolymerized in refluxing benzene using tetramethylguanidine (TMG) and other amine salts as catalysts (Fig. 1) to yield the poly(m-xylylenedisiloxanylenes), IIA - IIB. When the

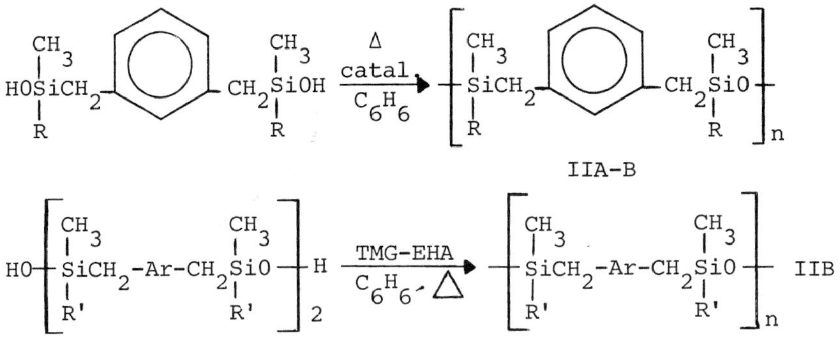

$$\underset{\substack{R}}{\overset{\substack{CH_3}}{HOSiCH_2}}\!\!-\!\!\bigcirc\!\!-\!\!CH_2\underset{\substack{R}}{\overset{\substack{CH_3}}{SiOH}} \xrightarrow[C_6H_6]{\overset{\Delta}{catal.}} \left[\underset{\substack{R}}{\overset{\substack{CH_3}}{SiCH_2}}\!\!-\!\!\bigcirc\!\!-\!\!CH_2\underset{\substack{R}}{\overset{\substack{CH_3}}{SiO}}\right]_n$$

IIA-B

$$HO\!\!\left[\underset{\substack{R'}}{\overset{\substack{CH_3}}{SiCH_2}}\!-Ar\!-CH_2\underset{\substack{R'}}{\overset{\substack{CH_3}}{SiO}}\right]_2\!\!H \xrightarrow[C_6H_6,\,\triangle]{TMG\text{-}EHA} \left[\underset{\substack{R'}}{\overset{\substack{CH_3}}{SiCH_2}}\!-Ar\!-CH_2\underset{\substack{R'}}{\overset{\substack{CH_3}}{SiO}}\right]_n \quad IIB$$

IIA R = CH_3 IIB R or R' = $CF_3CH_2CH_2$; Ar = $m\text{-}C_6H_4$

Fig. 1.   Syntheses of poly(m-xylylenedisiloxanylenes).

two bis-silanols, α,α'-bis(hydroxydimethylsilyl)- and α,α'-bis[hydroxy(3,3,3-trifluoropropylsilyl)-m-xylene, were co-polymerized with the appropriate bis(dimethylamino)silanes (Figure 2) or -disiloxanes (Figure 3), two families of various

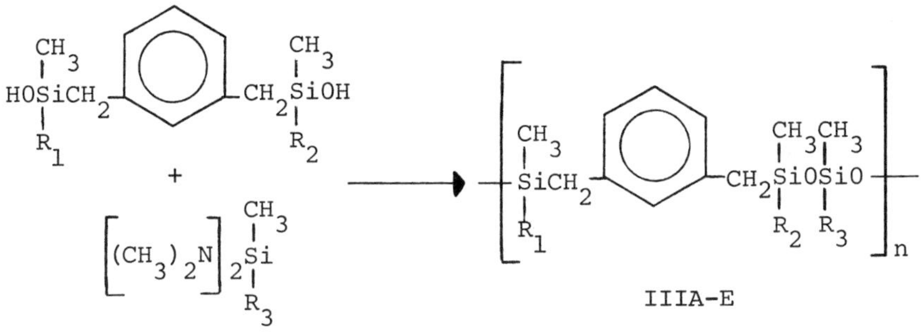

$$\underset{\substack{R_1}}{\overset{\substack{CH_3}}{HOSiCH_2}}\!\!-\!\!\bigcirc\!\!-\!\!CH_2\underset{\substack{R_2}}{\overset{\substack{CH_3}}{SiOH}}$$

+

$$\left[(CH_3)_2N\right]_2\underset{\substack{R_3}}{\overset{\substack{CH_3}}{Si}} \longrightarrow \left[\underset{\substack{R_1}}{\overset{\substack{CH_3}}{SiCH_2}}\!\!-\!\!\bigcirc\!\!-\!\!CH_2\underset{\substack{R_2\ \ R_3}}{\overset{\substack{CH_3\ CH_3}}{SiOSiO}}\right]_n$$

IIIA-E

IIIA $R_1 = R_2 = R_3 = R_4 = CH_3$ IIID $R_1 = R_2 = R_3 = R_f$

IIIB $R_1 = R_2 = CH_3$; $R_3 = R_f$ IIIE $R_1 = R_2 = R_f$; $R_3 = 97\% R_f$

IIIC $R_1 = R_2 = R_f$; $R_3 = CH_3$ $+ 3\% CH_2 = CH$

$$R_f = CF_3CH_2CH_2$$

Fig. 2.   Synthesis of poly(m-xylylenetrisiloxanylenes).

methyl- and 3,3,3-trifluoropropyl-substituted poly(m-xylylenetrisiloxanylenes), IIIA - IIIE, and -tetrasiloxanylenes, IVA - IVC, were obtained.

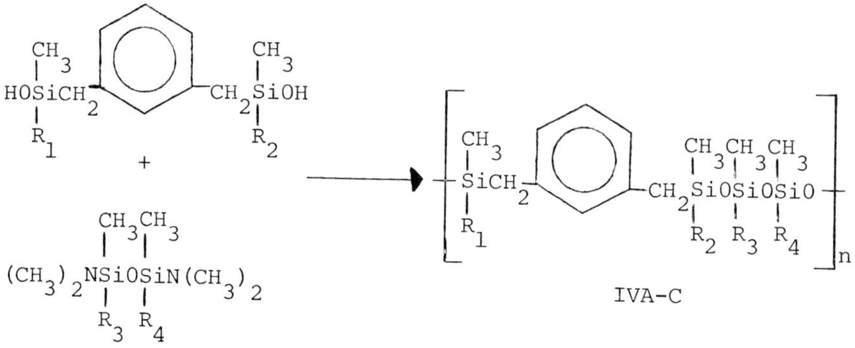

IVA $R_1 = R_2 = R_3 = R_4 = CH_3$ IVC $R_1 = R_2 = R_3 = R_4 = R_f$

IVB $R_1 = R_2 = CH_3$, $R_3 = R_4 = R_f$ $R_f = CF_3CH_2CH_2$

Fig. 3.   Synthesis of poly(m-xylylenetetrasiloxanylenes).

B.   Physical Properties

All of the synthesized polymers were characterized with respect to purity and chemical composition by conventional elemental and spectral analytical techniques. Viscosities, together with number average molecular weight data obtained by either vapor phase or membrane osmometry, are shown for the various polymers in Tables I - III. Glass-transition tempera-

TABLE I

Physical & Thermal Properties of Poly(m-xylylenedisiloxanylenes)

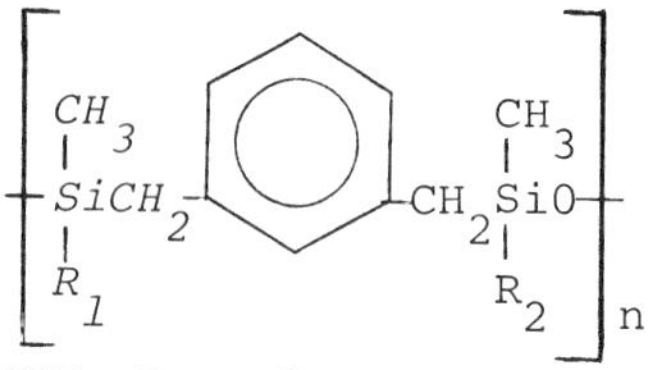

	$[\eta]$, dl/g 30°C	Mol Wt, $\overline{Mn}$	Tg, °C	TGA °C at % Wt Loss		
				10%	25%	50%
IIA, $R_1 = R_2 = CH_3$	0.458	38,880	-41	513	541	559
IIB, $R_1 = R_2 = R_f$	0.15		-19	436	461	482
IIC, $R_1 = R_2 = CH_3$; $(\underline{p}-C_6H_4)$	1.35 (inh)	55,100	-23	479	496	519

TABLE II

Physical & Thermal Properties of Poly(m-xylylenetrisiloxanylenes)

$$\left[\begin{array}{c}\underset{\underset{R_1}{|}}{\overset{\overset{CH_3}{|}}{-SiCH_2}}-\bigcirc-CH_2\underset{\underset{R_2}{|}}{\overset{\overset{CH_3}{|}}{Si}}\underset{\underset{R_3}{|}}{\overset{\overset{CH_3}{|}}{OSi}}O-\end{array}\right]_n$$

	$[\eta]$, dl/g 30°C	Mol Wt, $\overline{Mn}$	Tg, °C	TGA °C at % Wt Loss		
				10%	25%	50%
IIIA, $R_1 = R_2 = R_3 = CH_3$	0.355	14,500 (C_6H_6)	-62	537	557	573
IIIB, $R_1 = R_2 = CH_3; R_3 = R_f$	0.243	33,000 (C_6H_6)	-52	494	521	538
IIIC, $R_1 = R_2 = R_f; R_3 = CH_3$	0.11	21,000 (C_6H_6)	-44	453	472	493
IIID, $R_1 = R_2 = R_3 = R_f$	0.08		-35	450	471	491
IIIE, $R_1 = R_2 = R_f; R_3 =$ 97% R_f + 3% $CH_2 = CH$		21,000 (THF)	-34	441	471	493

TABLE III

Physical/Thermal Properties of Poly(m-xylylenetetrasiloxanylenes)

$$\left[\begin{array}{c}\underset{\underset{R_1}{|}}{\overset{\overset{CH_3}{|}}{-SiCH_2}}-\bigcirc-CH_2\underset{\underset{R_2}{|}}{\overset{\overset{CH_3}{|}}{Si}}\underset{\underset{R_3}{|}}{\overset{\overset{CH_3}{|}}{OSi}}\underset{\underset{R_4}{|}}{\overset{\overset{CH_3}{|}}{OSi}}O-\end{array}\right]_n$$

	ηinh dl/g 30°	Mol Wt, $\overline{Mn}$	Tg, °C	TGA °C at % Wt Loss		
				10%	25%	50%
IVA, $R_1 = R_2 = R_3 = R_4 = CH_3$		14,000 (C_6H_6)	-77	508	538	563
IVB, $R_1 = R_2 = CH_3; R_3 = R_4 = R_f$	0.08	13,000 (THF)	-59	478	504	529
IVC, $R_1 = R_2 = R_3 = R_4 = R_f$	0.142	22,000 (THF)	-38	446	473	499

tures were determined by means of differential scanning calorimetry (DSC), with $\Delta T = 20°C/min$ and the results are indicated in the three tables. Thermal characterization of the polymers

involved thermogravimetric analysis (TGA) under vacuum at
$\Delta T = 5°C/min$. Data obtained from the TGA curves for each of
the different polymers are shown in Tables I - III and will
be discussed in greater detail in an accompanying paper.[4] In
order to determine their resistance to hydrocarbon solvents,
polymer samples were first cured with di-t-butyl dicumyl peroxide at 170° and 2000 psi. The vulcanizates were then used
to measure volume swell ratios after immersion for 72 hours at
room temperature in hydrocarbons, such as isooctane and JP-4
jet fuel. Results obtained for a number of the polymers are
indicated in Table IV.

TABLE IV

Solvent Resistance of Cured m-Xylylenesiloxanylene Elastomers

(Cure Conditions: 70°/2000 psi/30 minutes)

| | | Volume Swell Ratio After 72 Hrs. at RT | |
Polymer	% Fluorine	Isooctane	JP-4
IIIA	0	221	200
IIIB	14.5	145	166
IIID	30.7	11	7.6
IIIE	27	27.7	24.2
IVA	0	170	150
IVB	21	60.5	77
IVC	32	0	4.5

C. Structure-Property Correlations

1. *Glass-Transition Temperature*

When the length of the siloxanylene moiety in poly(m-xylylenesiloxanylenes) is increased by one dimethylsiloxane unit,
e.g., from a disiloxanylene to a trisiloxanylene, the glass-
transition temperature is decreased by 15°C. Substitution of
one methyl group by a 3,3,3-trifluoropropyl group in a poly-
(m-xylylenesiloxanylene) increases the glass-transition temperature by 8 - 10°C.

2. *Thermal Behavior*

Structure-thermal stability relationships which can be derived for m-xylylenesiloxanylene and related phenylenesiloxanylene polymers from TGA data are more fully covered in the accompanying paper.[4] General conclusions, such as the marked decrease in stability with substitution of 3,3,3-trifluoropropyl for methyl groups as well as minimal change with minor variation in the siloxanylene moiety (i.e., from di- to tetrasiloxanylene) in permethylated polymers, can be readily drawn. Within the class of poly(m-xylylenesiloxanylenes), the polymers can be arranged by siloxanylene unit in a decreasing order of thermal stability as follows: 1,1,3,3-tetramethyldisiloxanylene > 1,1,3,5,5-pentamethyl-3,3,3-trifluoropropyltrisiloxanylene > 1,1,3,5,7,7-hexamethyl-3,5-bis(3,3,3-trifluoropropyl)tetrasiloxanylene > 1,3-dimethyl-1,3-bis(3,3,3-trifluoropropyl)disiloxanylene = 1,3,3,5-tetramethyl-1,3-bis-(3,3,3-trifluoropropyl)trisiloxanylene = 1,3,5-trimethyl-1,3,5-tris(trifluoropropyl)trisiloxanylene = 1,3,5,7-tetramethyl-1,3,5,7-tetrakis(3,3,3-trifluoropropyl)tetrasiloxanylene.

3. *Solvent Resistance*

In order for an elastomeric polymer to possess adequate hydrocarbon solvent or fuel resistance, a volume swell ratio of approximately 10% or less (as exhibited by its vulcanizate after immersion in the solvent) is considered a general prerequisite. In order to provide such volume swell ratios, a fluorine content of at least 30% by weight appears necessary in the polymer systems under investigation. Thus, in the poly(m-xylylenesiloxanylenes) only those polymers substituted with one methyl and one 3,3,3-trifluoropropyl group on each silicon atom can be expected to possess desired fuel resistance. The results obtained with regard to solvent resistance of these polymers, using vulcanizates, have been corroborated in more extensive but, as yet, incomplete studies on solvent interactions with m-xylylenesiloxanylene and related polymers. Using gas-liquid chromatography (GLC) to determine infinite-dilution activity coefficients, similar structure-property correlations with respect to fluorine content (and structure) of polymer and hydrocarbon insolubility were derived.[5]

III. EXPERIMENTAL

The two following examples are given to illustrate the general polymerization procedures used in the synthesis of the m-xylylenesiloxanylene polymers.

Poly[m-xylylene-1,3-dimethyl-1,3-bis(3,3,3-trifluoropropyl)-disiloxanylene]

In a 100-ml two-necked flask equipped with a magnetic stirrer, thermometer, and water trap filled with dry benzene and connected to a condenser with a drying tube ($CaSO_4$), were placed 10 grams of 1,3-bis[hydroxymethyl-(3,3,3-trifluoropropyl)silyl-m-xylylene], 5 ml of benzene and 0.11 grams of tetramethylguanidine 2-ethylhexoate. The mixture was heated under reflux for 1 hour at (80-90°) and some benzene removed to raise the temperature of the reaction mixture to 140°. The mixture was kept at that temperature for four hours and then heated at 160° and 0.2 torr for 21 hours. The gummy product obtained was dissolved in 35 ml of toluene, and the polymer precipitated with methanol (30 ml), washed twice with methanol (30 ml) and dried at 190° and 0.2 mm for 16 hours to yield 6.7 grams of poly[m-xylylene-1,3-dimethyl-1,3-bis(3,3,3-trifluoropropyl)-disiloxanylene]: translucent light brown gum; intrinsic viscosity (in toluene at 30°), 0.142 dl/g.

Poly[m-xylylene-1,1,3,5,5-pentamethyl-3-(3,3,3-trifluoropropyl)trisiloxanylene]

In a 100-ml three-necked flask, equipped with thermometer, stirrer, nitrogen inlet, and condenser which was connected to a solution of hydrochloric acid, were placed α,α'-bis(dimethyl-hydroxysilyl)-m-xylene (12.7 g, 0.05 mole), 10.5 grams of bis(dimethylamino)-3,3,3-trifluoropropylmethylsilane and 20 ml of toluene. The mixture was warmed to 110° and dimethylamine was evolved. After 15 minutes of heating, an additional 0.9 gram of bis(dimethylamino)-3,3,3-trifluoropropylmethylsilane was added dropwise over 5 minutes. The mixture was heated for 1.5 hours (total of 2 hours reaction time), hydrolyzed with 1.5 ml of water, and the water removed as an azeotrope. The polymer was dissolved in 50 ml of toluene, filtered, precipitated with 50 ml of methanol, washed twice with 30 ml of methanol, and dried at 170° and 0.1 torr for two hours to yield 15.4 grams of poly[m-xylylene-1,1,3,5,5-pentamethyl-3-(3,3,3-trifluoropropyl)trisiloxanylene]: translucent viscous gum; intrinsic viscosity (toluene at 30°), 0.243 dl/g.

IV. REFERENCES

1. Merker, R. L., and Scott, M. J., J. Polym. Sci., A, 2, 15 (1964).
2. Breed, L. W., Elliot, R. L., and Whitehead, M. E., J. Polym. Sci., A-1, 5, 2745 (1967).

3. Rosenberg, H., and Choe, E. W., <u>Abstracts of Papers</u>,
    <u>172nd Nat. Meeting</u>, <u>Amer. Chem. Soc.</u>, San Francisco, CA,
    Aug 1976, FLUO 41.
4. Goldfarb, I. J., Choe, E. W., and Rosenberg, H., <u>Coatings</u>
    <u>and Plastics Preprints</u>, <u>37</u> (1), 172 (1977).
5. Bonner, D. C., Chen, K. C., and Rosenberg, H., <u>Polymer</u>
    <u>Preprints</u>, <u>17</u> (2), 372 (1976).

THERMAL STABILITY-STRUCTURE CORRELATIONS IN ARYLENESILOXANYLENE POLYMERS

Ivan J. Goldfarb, Eui-won Choe* and Harold Rosenberg
Air Force Materials Laboratory

The need for new integral fuel tank sealants with use temperatures exceeding the limits of present materials has led to the synthesis in this laboratory of a family of new silicon-containing polymers. Inasmuch as the thermal stability of these polymers is a key factor in their subsequent utility as fuel tank sealants, a study of the effects of variation of their molecular structure on their resultant thermal stability was undertaken. The polymers studied were a series of arylenesiloxanylene polymers where the arylene linkage was either m-phenylene, p-phenylene or m-xylylene. The siloxanylene moieties varied from di- to tetrasiloxanylene with varying ratios of methyl and 3,3,3-trifluoropropyl groups. The principal experimental procedure for assessing thermal stability was dynamic thermogravimetry under vacuum at a programmed heating rate of 5°C per minute. The evolved gases during degradation of several polymers were also analyzed by combined TGA-mass spectrometry to ascertain the mode of thermal degradation. Thermo-oxidative degradation by small quantities of oxygen was determined by thermogravimetry under reduced air pressures (approx. 1 torr). The effects of the structural variations on the resulting thermal stability are discussed.

I. INTRODUCTION

Air Force requirements for integral fuel tank sealants with use temperatures exceeding the limits of present materials has led to the synthesis in this laboratory of a family of new silicon-containing polymers.[1] The polymers were a series of poly(arylenesiloxanylenes) where Ar is either m-phenylene or m-xylylene, R_1, R_2 and R_3 are either methyl or 3,3,3-trifluoropropyl, and x = 0, 1 or 2.

---

*Present address: Celanese Research Company, Summit, NJ 07901

$$
\left[\begin{array}{c} CH_3 \\ | \\ -Si - Ar - \\ | \\ R_1 \end{array} \begin{array}{c} CH_3 \\ | \\ Si - 0 \\ | \\ R_2 \end{array} \left(\begin{array}{c} CH_3 \\ | \\ Si - 0 \\ | \\ R_3 \end{array} \right)_x \right]_n
$$

Inasmuch as the thermal stability of these polymers is a key factor in their subsequent utility as fuel tank sealants, an understanding of the effects of variation of molecular structure on their resultant thermal stability was necessary. It has been known that polydimethylsiloxane decomposes and reverts into low molecular weight cyclic trimers and tetramers upon heating at 300°C but it has been reported that siloxane polymer stability could be improved by incorporation of arylene segments between two silicon atoms.[2] This paper describes our investigations on the effects of the variation of the arylene moiety, the length of the siloxane linkage and the substitution of fluorinated groups for methyl groups.

II. EXPERIMENTAL

Polymer samples were synthesized in this laboratory by procedures described in the previous paper.[1] Polydimethylsiloxane was prepared by the ring-opening reaction of octamethylcyclotetrasiloxane with tetramethylammonium hydroxide and had a molecular weight, as determined by its intrinsic viscosity relationship, of 172,000. Thermogravimetric analysis (TGA) was performed on an Ainsworth semi-micro automatic balance which was modified to provide direct temperature measurement in the polymer sample. This was accomplished by utilizing thermocouple wires as hangdown wires for the balance with an electrical takeoff which did not disturb the weighing.[3] An automatic data acquisition system was utilized to record and reduce the raw data. All TGA runs were performed at a programmed heating rate of 5°C per minute and 10^{-5} torr pressure with the exception of the several runs performed at 1 torr air to determine the polymer oxidation sensitivity. Mass spectral thermal analysis was accomplished utilizing a Finnigan quadrupole mass spectrometer installed directly into a vacuum system housing a Cahn RH electrobalance. The mass spectral data as well as weight and temperature were automatically digitized and recorded on magnetic tape for subsequent computer reduction and analysis on a CDC 6600 computer.

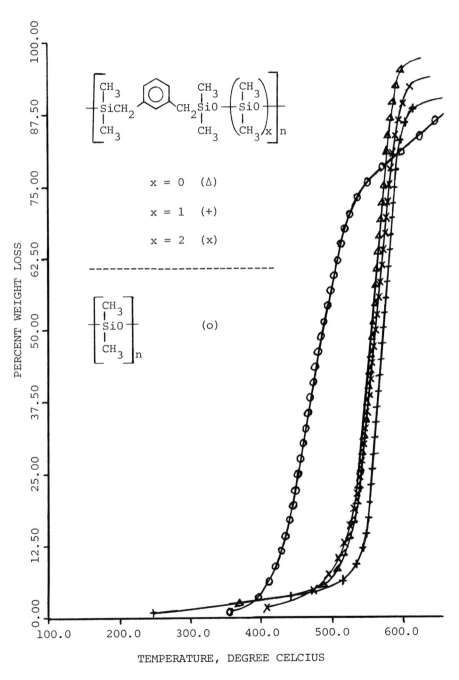

Fig. 1.   Thermogravimetric analyses of polydimethylsilox-ane and poly(m̲-xylylenesiloxanylenes).

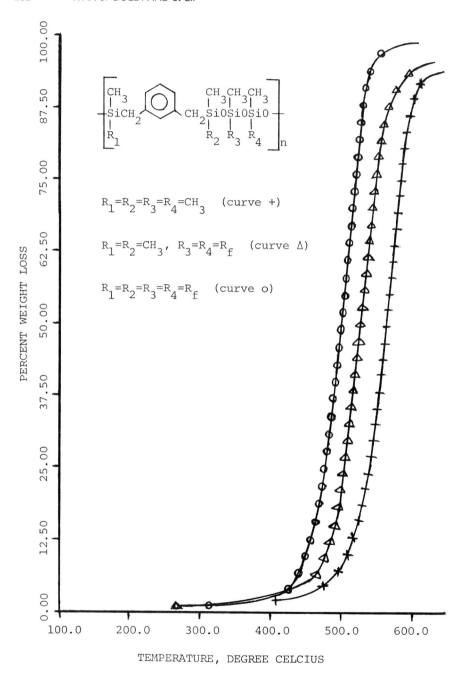

Fig. 2. *Thermogravimetric analyses of poly(m̠-xylylene-tetrasiloxanylenes).*

III. RESULTS AND DISCUSSION

The TGA curves for polydimethylsiloxane and all-methyl-ated poly($\underline{m}$-xylylenesiloxanylene) polymers with the general

$$\left[\begin{array}{c} CH_3 \\ | \\ -Si - CH_2 \!\!\!\bigcirc\!\!\!\!-\!\!\!CH_2 - Si - O \!\!\left(Si - O \right)_{\!\!x} \\ | \qquad\qquad\qquad | \qquad | \\ CH_3 \qquad\qquad CH_3 \quad CH_3 \end{array} \right]_n$$

structure where x = 0, 1 and 2 are shown in Figure 1. It can be seen that the hexamethyltrisiloxanylene polymer showed higher thermal stability than those of polymers with tetra-methyldisiloxanylene and octamethyltetrasiloxanylene moie-ties, and significantly higher than that of the polydimethyl-siloxane shown for comparison. The effects of variation of numbers and positions of 3,3,3-trifluoropropyl groups on the silicon atoms for poly($\underline{m}$-xylylenetetrasiloxanylene) are shown in Figure 2. It is obvious that substitution of fluorinated groups for methyl groups decreases the thermal stability of the polymers.

TABLE 1

Thermal Decomposition Temperatures ($^\circ$C)
of Arylenesiloxanylene Polymers

No.	R_1	Ar	R_2	R_3	R_4	T_{25}
1.	Me	Xy	Me	–	–	540
2.	Me	Xy	Me	Me	–	557
3.	Me	Xy	Me	R_f	–	521
4.	R_f	Xy	R_f	Me	–	472
5.	R_f	Ph	R_f	Me	–	441
6.	R_f	Xy	R_f	R_f	–	471
7.	Me	Xy	Me	Me	Me	538
8.	Me	Xy	Me	R_f	R_f	504
9.	Me	Ph	Me	R_f	R_f	482
10.	R_f	Xy	R_f	R_f	R_f	473
11.	R_f	Ph	R_f	R_f	R_f	450

Me = CH_3- , R_f = $CF_3CH_2CH_2-$, Xy = $-CH_2-\underline{m}-C_6H_4CH_2-$, Ph = $-\underline{m}-C_6H_4-$

It is convenient to choose a single figure of merit for the thermal stability of these polymers and, inasmuch as they all gave TGA curves of similar shape and left practically no residue upon complete degradation, the temperature at which 25 percent weight loss occurred was chosen. This weight loss is sufficiently high so that spurious effects of early weight loss could be avoided and still early enough in the reaction that a measure of the initial decomposition could be assumed. These values are shown for the various polymers studied in Table 1. The columns R_1, R_2, R_3 and R_4 refer to the substituents on the silicon atoms in the structural formula shown at the beginning of this paper and Ar refers to the particular arylene moiety incorporated into the backbone of the polymer. It should be noted that the disiloxanylene polymer has no R_3 or R_4 groups while the trisiloxanylenes have no R_4 group. From Table 1 a number of generalizations can be made:

1. Both in the all-methyl-substituted polymers (polymers No. 1, 2 and 7) and in those polymers with one fluoroalkyl group on every silicon atom (polymers No. 6 and 10), change in the length of the siloxane linkage had an insignificant effect on the thermal stability.

2. Replacing methyl groups by trifluoropropyl groups in positions not adjacent to the arylene moiety (polymers No. 3, 8 and 9) causes a reduction in thermal stability (as evidenced by the T_{25}) of approximately 35° while a similar replacement at sites adjacent to the arylene moiety (polymers No. 4, 5, 10 and 11) causes double that reduction.

3. The total number of methyl groups replaced by fluoroalkyl groups is not important, only their placement.

4. The m-phenylene-containing polymer has a thermal stability 20° to 30° lower than its m-xylylene analog.

Although the principal aim of this study was thermal degradation and not oxidation of polymers, it was noted that some of these polymers seemed to have a sensitivity towards trace quantities of air at elevated temperatures. The effects of small quantities of air were therefore studied for several of the polymers. TGA runs were made as before except with 1 torr of air present. Polymer No. 10 in Table 1 exhibited a T_{25} of 473°C when run under 10^{-5} torr vacuum but this was reduced to 450°C at 1 torr. Polymer No. 11 showed no reduction at all when run at 1 torr. Therefore, it would appear that the m-phenylene linkage, while less stable thermally, is less susceptible to trace oxidation.

In order to obtain some information on the mode of decomposition of these polymers, several (polymers No. 1, 6 and 7 in Table 1) were analyzed by combined TGA-MASS spectrometry. In all cases there was little or no temperature resolution in the mass spectral scans indicating that all the volatile products were evolved in a single reaction. In the scans of all three polymers fragment ions were detected typical of methylsiloxanes, e.g., m/e 73 $[(CH_3)_3Si^+]$. Polymer No. 1, the disiloxanylene, produced peaks at m/e 221, 207 and 193 indicating loss of CH_2 from the monomer unit. There was a narrow temperature range of decomposition peaking at approximately 565°C. This is in good agreement with the TGA reported earlier. Polymer No. 7, a tetrasiloxanylene, produced a spectrum with many peaks similar to that of octamethylcyclotetrasiloxane indicating the breakdown of the siloxanylene portion of the polymer. Polymer No. 6, the trifluorinated trisiloxanylene, exhibited a spectrum which was remarkably similar to the previous ones considering the presence of the fluoroalkyl groups. A sizeable peak at m/e 51 indicated CF_2H^+ was present but the most noteworthy feature of the spectrum was the large intensity of ions relating to alkylbenzenes. Apparently the presence of fluoroalkyl groups allows the xylylene portion to be preferentially evolved as gaseous reaction products far more easily than with the completely methyl-substituted polymers.

IV. ACKNOWLEDGEMENT

The authors wish to thank Mr. Gary Doll for experimental assistance with the TGA determinations and Dr. E. Grant Jones and Mr. Paul Benadum for the mass spectral analysis.

V. REFERENCES

1. Rosenberg, H. and Choe, E. W., Coatings and Plastics Preprints, 37 (1), 166 (1977).

2. Breed, L. W., Elliot, R. L. and Whitehead, M. E., J. Poly. Sci., A-1, 5, 2745 (1967).

3. Goldfarb, I. J., Bain, D. R., McGuchan, R. and Meeks, A. C., Polymer Preprints, 12 (1), 733 (1971).

POLYPHOSPHAZENES: STRUCTURE AND APPLICATIONS

Robert E. Singler and Gary L. Hagnauer
U.S. Army Materials and Mechanics Research Center

During the past ten years it has been conclusively demonstrated that stable polyorganophosphazenes can be prepared from an inorganic precursor, polydichlorophosphazene, by careful control of polymerization and substitution reaction conditions.   A wide variety of polymers have been prepared by this route. Molecular weights of these polymers range from 10^5 to 10^6. Evidence accumulated so far indicates that these polymers can be prepared free of gel and extensive chain branching.   Both semicrystalline and amorphous polymers have been prepared. The successful preparation of stable polyorganophosphazenes appears to have resulted in a new class of inorganic backbone polymers for both research and development.   This paper describes some of the dilute solution characterization and technological development of these polymers.

I.   INTRODUCTION

The study of open-chain polyphosphazenes has attracted increasing attention in the past 10 years both from the viewpoint of fundamental research and technological development (1, 2, 3).   Interest has centered around investigation of synthetic pathways and fundamental properties and the development of these polymers for certain rubber and plastic applications.

Prior to 1965 there was only minor interest in the phosphazene high polymers, primarily because as a class they were viewed to be intractable or hydrolytically unstable materials. The breakthrough came with the work of Allcock and coworkers (4-5) describing the synthesis of stable polyorganophosphazenes which can be prepared by the melt ring-opening polymerization of hexachlorocyclotriphosphazene (trimer) followed by the reaction of soluble polydichlorophosphazene with various organo-nucleophiles as shown below.

Care must be taken to avoid the formation of insoluble polydichlorophosphazene, generally known as "inorganic rubber", and in the selection of proper conditions for conversion into polyorganophosphazenes. Although problems still exist in the synthetic area, work has progressed to where a wide variety of polyorganophosphazenes are now available (1, 2, 6-10). This paper will deal primarily with the dilute solution properties and potential applications for some of these polymers.

II.   DILUTE SOLUTION CHARACTERIZATION

A.   Experimental

Standard dilute solution techniques were used to characterize the polymers with tetrahydrofuran (THF) as the solvent at 25°C. Cannon-Ubbelohde dilution viscometers were employed for intrinsic viscosity $[\eta]$ determinations. The temperature was controlled at ±0.02°C, and solvent efflux times were sufficiently long to justify neglecting kinetic energy corrections. A FICA 50 light scattering instrument (λ_0 = 5461 Å) was used to evaluate weight average molecular weights $\bar{M}_w$, second virial coefficients A_2, and z-average radii of gyration R_g. The instrument was calibrated with benzene (R_B=16.3 x 10^{-6}) and scattered light intensities were measured at preset angles between 30° and 150°. Refractive index increments (dn/dC) were determined using a Brice-Phoenix differential refractometer.

A Waters ANAPREP gel permeation chromomatograph with refractive index monitor was used for the GPC analyses. The column set consisted of 4-ft x 3/8-in styragel columns with porosity ratings designated as 5 x 10^6, 10^6, (1.5-7) x 10^5,

10^5, 10^4 and 10^3 Å for the chromatograms in Figs. 1 and 2, and 5 x 10^6, 10^6, (1.5-7) x 10^5, 10^4 Å for the chromatograms in Fig. 3. A flow rate of 1 ml/min was maintained with THF as the eluent at 35°C. Elution volume is designated in "counts" where 2.5-ml of effluent as measured by a syphon device is 1 count. Calibration above M = 7.1 x 10^6 was achieved by linear extrapolation of data obtained from narrow distribution polystyrene standards. The molecular weight (MW) scale in Figs. 1 and 2 are based on the polystyrene calibration and does not represent actual MWs for the polyphosphazenes.

A crude fractionation of $[(C_6H_5O)(4-ClC_6H_4O)PN]_n$ was achieved by precipitation from dilute THF solution with methanol as the nonsolvent. Weight percentages of the recovered fractions are indicated in Table 1.

B. Macromolecular Structure

Polyphosphazenes are found to have characteristically high molecular weights and broad molecular weight distributions (MWD) (2, 11-13). Average molecular weights, molecular weight distribution, chain structure and solubility characteristics depend on polymerization and substitution reaction conditions as well as the functionality of the polymer chain side group. Polydichlorophosphazene is hydrolytically unstable and therefore difficult to characterize by dilute solution techniques. Careful handling and chloride replacement with trifluoroethoxide stabilizes the polymer and tends to preserve the structure of the polyphosphazene backbone. A gel permeation chromatogram of polydichlorophosphazene would be similar to that shown for $[(CF_3CH_2O)_2PN]_n$ in Fig. 1. Typically, $[(CF_3CH_2O)_2PN]_n$ polymers and therefore presumably polydichlorophosphazene, have a broad distribution with high molecular weight tail. Fractionation and dilute solution studies suggest the presence of long chain branching in some $[(CF_3CH_2O)_2PN]_n$ samples.

In the case of aryloxide substitution, complete chloride replacement is difficult to achieve if the polymers are branched. However, at elevated substitution reaction temperatures (T > 125°C), it is possible to completely substitute not only branched but also lightly crosslinked polydichlorophosphazene (14). In the substitution process, branch and crosslink junctions in the chain may undergo scission yielding fully substituted, linear polyaryloxyphosphazenes. The chromatogram in Fig. 1 for $[(4-CH_3C_6H_4O)_2PN]_n$ is typical for the polyaryloxyphosphazenes.

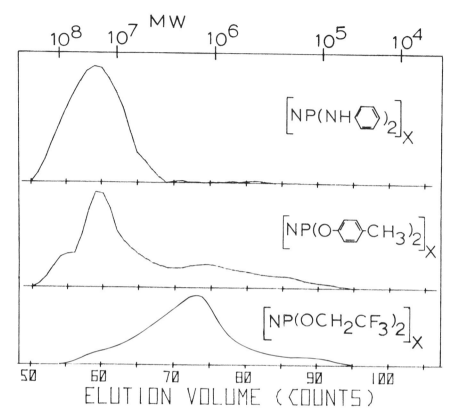

Fig. 1.   Gel Permeation Chromatograms of Polyorganophosphazenes

The unusually high MW indicated for polyanilinophospha-
zene (Fig. 1) may be the consequence of a large hydrodynamic
volume due to an extended chain structure or perhaps is the
result of chain association in solution.

The polymers were prepared from different batches of
polydichlorophosphazene using different reaction conditions.
Since the chemical structures, and perhaps the chain structure,
of the polymers were different, comparisons of dilute solution
properties should be made with caution.   The dilute solution
data for the polymers in Fig. 1 are given in Table 1.

TABLE 1

Dilute Solution Characterization

Sample	$[\eta]$ (dl/g)	dn/dC (ml/g)	$M_w \times 10^{-6}$	$A_2 \times 10^4$ (ml-mole/g^2)	R_g (Å)
$[(C_6H_5NH)_2PN]_n$	1.43	0.145	4.68	0.681	660
$[(4-CH_3C_6H_4O)_2PN]_n$	2.13	0.164	2.2	1.9	650
$[(CF_3CH_2O)_2PN]_n$	1.27	-0.027	1.5	0.68	306
$[(4-CH_3CH_2C_6H_4O)-(C_6H_5O)PN]_n$					
A	3.2	-	-	-	-
B	2.5	-	-	-	-
C	1.7	-	-	-	-
$[(4-ClC_6H_4O)-(C_6H_5O)PN]_n$					
unfractionated	2.37	0.157	2.74	1.9	855
(8.5%) fraction 1	3.72	0.157	10.3	1.7	1650
(58.4%) fraction 2	2.56	0.154	3.71	1.6	958
(26.9%) fraction 3	1.85	0.153	1.91	1.8	709
(6.2%) fraction 4	0.849	0.146	.890	9.0	576
$[(C_6H_5O)_2PN]_n$	1.74	0.169	2.40	2.0	800
$[(4-ClC_6H_4O)_2PN]_n$	1.70	0.153	1.09	5.1	766

Chromatograms of the phosphazene copolymer $[(C_6H_5O)-(4-CH_3CH_2C_6H_4O)PN]_n$ are illustrated in Fig. 2. Samples A and B were prepared in different laboratories by bulk polymerization from different batches of hexachlorocyclotriphosphazene. The somewhat different MWD observed for these samples may be a result of differences in trimer purity. Sample C was prepared by solution polymerization in 1,2,4-trichlorobenzene at 200°C, with polyphosphoric acid (PPA, 1%) as a catalyst (15). The MW of the sample prepared by solution polymerization is smaller than for the bulk polymerized polymers, and unexpectedly, Sample C has a much broader MWD. The reasons for broad distribution in C are not understood. The intrinsic viscosities for the polymers in Fig. 2 are given in Table 1.

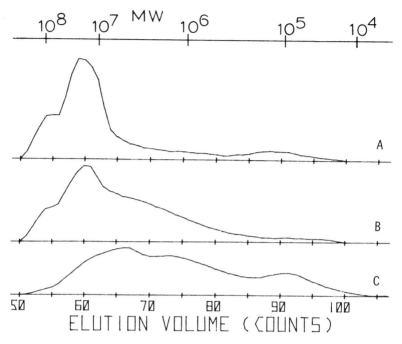

Fig. 2. Gel Permeation Chromatograms of $[(C_6H_5O)-$
$(4-CH_3CH_2C_6H_4O)PN]_n$

Another copolymer $[(C_6H_5O)(4-ClC_6H_4O)PN]_n$ was fractionated by precipitation with a nonsolvent in an attempt to determine whether factors other than MW, such as branching or chemical heterogeneity in polymer structure, may be contributing to the separation. Fractions 1-4 are ranked in order of increasing solubility in the solvent/nonsolvent mixtures. Homopolymers incorporating the side groups of the copolymer were also characterized. Both the GPC (Fig. 3) and dilute solution data (Table 1) indicate that separation by MW and molecular size occurred. It is also noted that the (dn/dC) value for the copolymer is intermediate to the respective homopolymer values. However, (dn/dC) seems to decrease with decreasing MW of the fraction until the value for fraction 4 is actually less than the value for either homopolymer. Furthermore, the value for A_2 is considerably larger for fraction 4. As shown in Fig. 3, the apparent low MW component in fraction 4 may contribute to the observed differences. The effects of chain ends become more important with decreasing molecular weight. Narrower fractions and spectroscopic studies are required to determine whether chain branching or chemical heterogeneities, such as block structure or hydrolysis of the chain backbone, are present in the polymer.

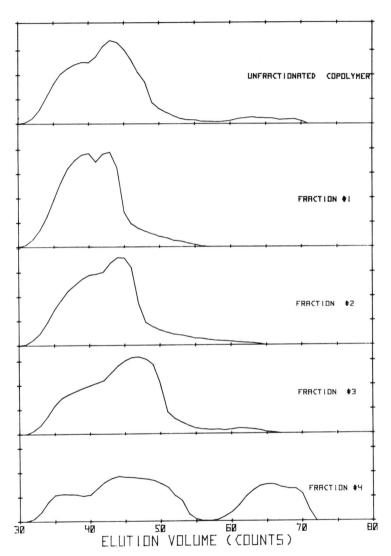

Fig. 3. *Gel Permeation Chromatograms of* $[(C_6H_5O)-$
$(4\text{-}ClC_6H_5O)PN]_n$ *Copolymer*

III. APPLICATIONS

There is a continuing search for new elastomeric materials which will provide service over a wide temperature range (-70° to over 400°F) and for improved flame retardant elastomers and plastics. Conventional organic polymers are somewhat limited in these respects. In contrast, purely inorganic materials, while having greater thermal stability and flame retardancy, generally do not have the physical properties such as flexibility or elasticity necessary for these applications. The question then arises as to whether one can find a suitable compromise between these two classes of materials. The inorganic based polyorganosiloxanes are the best known example of this compromise. These same questions and requirements form the basis for much of the recent technological interest in the polyphosphazenes.

A. Fluoroelastomers

Much of the current technological interest in the polyorganophosphazenes is based on a modification of the substitution reaction outlined above. Rose (16) first prepared phosphazene "hybrid copolymers" by using a mixture of fluoroalkoxides during the substitution reaction:

$$\left\{ \begin{array}{c} Cl \\ | \\ P = N \\ | \\ Cl \end{array} \right\}_n \quad \xrightarrow[R_{f'}CH_2ONa]{R_fCH_2ONa} \quad \left\{ \begin{array}{c} OCH_2R_f \\ | \\ P = N \\ | \\ OCH_2R_{f'} \end{array} \right\}_n \qquad \begin{array}{l} R_f = CF_3 \\ \\ R_{f'} = n\text{-}C_3F_7 \end{array}$$

Whereas the poly(alkoxy- or aryloxyphosphazene) homopolymers generally are flexible film forming thermoplastics, the copolymers and terpolymers are often elastomers. Under proper conditions the substitution step probably gives a random distribution of different polymer repeat units as illustrated below:

$$\sim \begin{array}{c} OCH_2R_f \\ | \\ P = N \\ | \\ OCH_2R_f \end{array} - \begin{array}{c} OCH_2R_f \\ | \\ P = N \\ | \\ OCH_2R_{f'} \end{array} - \begin{array}{c} OCH_2R_{f'} \\ | \\ P = N \\ | \\ OCH_2R_{f'} \end{array} \sim$$

Several phosphazene fluoroelastomer copolymers were subsequently prepared and evaluated (17). These copolymers had low glass transition temperatures (T_g -64° to -77°C), good thermal stability, and were extremely solvent resistant. Variations in the basic copolymer structure are now employed to enhance properties for specific applications. For example, a small amount ($\sim$ 1%) of a reactive pendant group can be added along with the fluoroalkoxides in the substitution step to provide more reactive curing sites. These terpolymers, based largely on the $[(CF_3CH_2O)(HCF_2C_3F_6CH_2O)PN]_n$ copolymer, are under development by both private industry and government laboratories.

The physical properties of the phosphazene fluoroelastomer terpolymers appear very promising (2,3). Vulcanization can be achieved with both organic peroxides and sulfur curing systems. Vulcanizates can be reinforced with silicas, carbon blacks and clays to give stocks with a wide range of properties. Tensile strengths over 2500 psi with elongations of 200% have been obtained. However, more typically, vulcanizates with tensile strengths of approximately 1500 psi and elongations of 100-200% are prepared, along with a suitable balance of low temperature flexibility, heat and fluid resistance, low compression set, and other properties for intended applications.

Because of the high fluorine content the phosphazene fluoroelastomers are relatively expensive and will be viewed primarily as speciality elastomers for extreme service requirements of temperature and solvent resistance. Thus, they will be compared with other solvent resistant speciality rubbers (3, 18-20). The potential applications include items such as O-rings for hydraulic systems, oil seals, fuel hoses, coupling gaskets, coated fabrics, vibration damping equipment, and obturator pads. Advanced development is in progress in several of these areas such as arctic fuel handling applications (21).

B. Flame Retardant Polymers

An area where the polyorganophosphazenes are expected to have an even greater potential is for flame retardant applications. Most work to date has been with the polyaryloxyphosphazenes with the general structures $[(ArO)_2PN]_n$ and $[(ArO)-(Ar'O)PN]_n$. Whereas most organic polymers either burn rapidly in air or evolve excessive amounts of smoke and toxic gases, recent studies have shown the polyaryloxyphosphazenes to be self extinguishing in air with limiting oxygen index values (LOI) of 24-65 and only moderate smoke evolution (22, 23). Some representative data are shown in Table 2.

TABLE 2

Flammability Comparisons - Polyaryloxyphosphazenes
and Reference Polymers*

Polymer	LOI (ASTM D2863)	$D_m{}^+$ (Flaming Mode)
Polyethylene	17	150
Polystyrene	18	468
Silicone Rubber	26	385
Polycarbonate	27	>660
Polyvinylchloride	44	525
$[(RC_6H_4O)_2PN]_n$		
R = $4-\eta C_4H_9O$	24	60
$4-C_2H_5$	25	305
$4-CH_3O$	26	120
$3-CH_3$	28	–
$4-CH_3$	27	321
H	33	343
$4-Cl$	44	455
$4-Br$	65	–
$[(RC_6H_4O)(R'C_6H_4O)PN]_n$		
R = H, R' = $4-C_2H_5$	27	331
R = H, R' = $4-CH_3O$	25	151
R = $4-CH_3O$, R' = $4-C_2H_5$	24	139
R = $4-CH_3O$, R' = $4-sec-C_4H_9$	26	154

* Ref (22, 23)

+ NBS Smoke Density Chamber, Maximum specific optical density

The smoke evolution for the polyaryloxyphosphazenes can
be reduced even further by selection of the proper choice of
inert fillers. Quinn and Dieck (23) have shown that blending
a series of nonhalogenated homopolymers with various amounts
of three common fillers, alumina trihydrate, hydrated silica,
and calcium carbonate significantly increased oxygen index
and lowered smoke density values over those shown in Table 2.
Some representative data using alumina trihydrate are shown
in Table 3.

TABLE 3

Flame and Smoke Properties of Filled Polyaryloxyphosphazenes[*]

Additive (phr)[+]	$[(C_6H_5O)_2PN]_n$	
	LOI (ASTM D2863)	D_m (Flaming Mode)[#]
0	33.9	322
10	31.0	282
25	34.8	304
50	38.3	211
	$[(4-CH_3OC_6H_4O)_2PN]_n$	
0	25.5	120
10	35.3	165
25	35.3	137
50	43.5	73

[*] Ref 23
[+] Additive (alumina Trihydrate) used per one hundred parts of polymer
[#] NBS Smoke Density Chamber, maximum specific optical density

Low smoke, flame retardant foams and wire coverings have been prepared using polyaryloxyphosphazene copolymers (24-26, 15). In the preparation of the closed cell foams, alumina trihydrate is employed as the flame retardant filler and azobisformamide provides the basis for the chemical expansion process. The polymer is crosslinked with a combination of peroxides before and during the chemical blowing step. Foam densities from 4 to 10 lb/ft^3 can be obtained along with low NBS smoke densities (D_m flaming <100) and other outstanding flame retardant properties. The physical and mechanical properties appear to be suitable for certain military and commercial applications. Although detailed studies on the toxicological effects of the combustion products have not been reported, indications are that the polyaryloxyphosphazenes are much less hazardous than current foam and wire covering materials used in flame retardant applications (24).

DiEdwardo (27) has recently shown the efficacy of blending polyphosphazenes with commercial polymeric materials to reduce their flammability. Blending 10 wt% aryloxy and anilino poly-phosphazenes with polyester, polystyrene, and acetate polymers significantly increased the oxygen index values and made the polymers char rather than drip upon burning. The polyanilino-phosphazenes may be especially advantageous in these respects.

IV   CONCLUSION

Polyphosphazene chemistry has made remarkable strides during the past 10 to 12 years, and the interest is expected to continue. Work leading to a more thorough understanding of the factors controlling the preparation, chain structure and other fundamental properties of these polymers will continue and ultimately aid in their technological development.

Because of the variety of substitution reactions that can be carried out with polydichlorophosphazene, a large number of polyorganophosphazenes can be envisioned which may have potential for development. Although much of the future work will be in the areas of speciality elastomers and flame retardant polymers, other areas appear promising. For instance, certain polyphosphazenes may have utility as biomedical replacement materials, such as artificial heart valves. Certain water soluble polyaminophosphazenes may prove useful as chemotherapeutic drug carrier molecules (1), and this aspect will be the subject of a subsequent paper in this symposium.

V   ACKNOWLEDGEMENTS

The authors wish to acknowledge Mr. Richard Matton and Mr. Charles Goddard for the preparation of some of the polymers described in this paper. We also wish to thank Carolyn Shimansky for her patience and attention to details during the preparation of this manuscript.

VI   REFERENCES

1. Allcock, H. R., "Phosphorus-Nitrogen Compounds", Academic Press, New York (1972); Allcock, H. R., Angew. Chem. Int. Ed. Engl. 16, 147 (1977).
2. Singler, R. E., Schneider, N. S., Hagnauer, G. L., Polym. Eng. & Sci., 15, 321 (1975).
3. Tate, D. P., J. Polym. Sci., C48, 33 (1974).
4. Allcock, H. R., Kugel, R. L., and Valan, K. J., Inorg. Chem. 5, 1709 (1966).
5. Allcock, H. R., and Kugel, R. L., ibid., 5, 1716 (1966).
6. Allcock, H. R., and Moore, G. Y., Macromolecules 8, 377 (1975).
7. White, J. E., Singler, R. E., and Leone, S. A., J. Polym. Sci., Polym. Chem. Ed. 13, 2531 (1975).
8. Dieck, R. L., and Goldfarb, L., ibid., 15, 361 (1977).
9. White, J. E., and Singler, R. E., ibid., 15, 1169 (1977).
10. Busulini, L., Osellame, M., Lora, S., and Pezzin, G., Macromol. Chem. 178, 277 (1977).

11. Hagnauer, G. L., and LaLiberte, B. R., *J. Polym. Sci.*, *Polym. Phys. Ed.*, 14, 367 (1976).
12. Carlson, D. W., O'Rourke, E., Valaitis, J. K., and Altenau, A. G., *J. Polym. Sci. Polym. Chem. Ed.*, 14, 1379 (1976).
13. Hagnauer, G. L., and LaLiberte, B. R., *J. Appl. Polym. Sci.*, 20, 3073 (1976).
14. LaLiberte, B. R., and Hagnauer, G. L., "Nucleophilic Substitution of Cyclic and Polymeric Dichlorophospha-zenes with m-chlorophenoxide", AMMRC TR 76-17, June 1976 (AD A027368).
15. Thompson, J. E., Wittman, J. W., and Reynard, K. A., "Open-Cell Fire Resistant Foam", Horizons, Inc., Cleveland, Ohio, NASA Contract NAS9-14717, April 1976 (N76-27424).
16. Rose, S. H., *J. Polym. Sci., Part B* 6, 837 (1968).
17. Rose, S. H., Reynard, K. A., *Polym. Prep. Am. Chem. Soc. Div. Polym. Chem.*, 13(2), 778 (1972).
18. Kyker, G. S., and Antowiak, T. A., *Rubber Chem. Tech.*, 47, 32 (1974).
19. Tate, D. P., *Rubber World*, 172, 41 (1975).
20. Touchet, P. and Gatza, P. E., *J. Elastomers Plast.*, 9, 8 (1977).
21. Antowiak, T. A., and Welvaert, D. M., "Fabrication of a Low Temperature Fuel Hose from Phosphonitrilic Fluoroelastomer", Firestone Tire & Rubber Co., Akron, Ohio, U.S.A. MERADCOM Contract DAAG53-75-C-0187, November 1976 (AD-A036903).
22. Reynard, K. A., Gerber, A. H., and Rose, S. H., "Synthesis of Phosphonitrilic Elastomers and Plastics for Marine Application". Horizons, Inc., Cleveland, Ohio, AMMRC CTR72-29, December 1972 (AD755 188).
23. Quinn, E. J., and Dieck, R. L., *J. Fire & Flammability*, 7, 5 (1976); *ibid*, 358 (1976).
24. Reynard, K. A., Sicka, R. W., and Thompson, J. E., "Poly(aryloxyphosphazene) Foams", Horizons, Inc. Cleveland, Ohio, Naval Ship Systems Contract N0002473C5474, June 1974 (AD-A009425).
25. Quinn, E. J., and Dieck, R. L., *J. Cell Plastics*, 13, 96(1977).
26. Reynard, K. A., and Vicic, J. A., "Polyphosphazene Wire Coverings", Horizons, Inc., Cleveland, Ohio, Naval Ship Engineering Center Contract N00024-75-C-4402.
27. DiEdwardo, A. H., Zitomer, F., Steutz, D., Singler, R. E., and Macaione, D. P., *Organic Coatings Preprints* 36(2), 737 (1976).

TRANSITION TO THE MESOMORPHIC STATE IN POLYPHOSPHAZENES

N. S. Schneider, C. R. Desper, R. E. Singler
M. N. Alexander, and P. L. Sagalyn
U. S. Army Materials and Mechanics Research Center

The poly(organophosphazenes) of general structure $[(RO)_2PN]_n$, where RO may be an alkoxy or an aryloxy group, are generally crystalline and exhibit both a first order transition to a mesomorphic phase at a temperature designated T(1) and the transition to the isotropic melt at a temperature T_M which is 150 to 250°C higher. An investigation of the behavior at T(1) and T_M by scanning calorimetric and thermal expansion methods is presented. It is shown that the thermal history of the sample at temperatures above T(1) has a strong influence on the endotherm at T(1) implying that it is possible to alter the structure of the mesomorphic phase and, in turn, the crystalline structure formed on cooling. The results of x-ray diffraction studies indicate that the T(1) transition involves a transformation from the crystalline structure to a metrically hexagonal lattice exhibiting lateral order but longitudinal disorder and that side chain packing controls the interchain distance. Preliminary broad line NMR data show that the disorder above T(1) is dynamic rather than positional and arises from the onset of rapid side chain and backbone motions.

I. INTRODUCTION

A large variety of semicrystalline polyphosphazenes have now been prepared (1, 2, 3) by the procedure first outlined by Allcock. These polymers have the general formula $[(RO)_2PN]_n$ where RO may be either an alkoxy or an aryloxy group. One of the distinctive aspects of the thermal transition behavior of these materials is the fact that for many of the polyphospha- zenes two first order transitions can be detected, separated by as much as 150 to 250°C (2, 3). The lower transition, referred to as T(1), represents the transformation from a crystalline to a mesormorphic state, while the upper transition T_M represents passage to the isotropic melt. In this paper we consider evidence on the behavior and nature of the transitions at T(1) and T_M as revealed by calorimetric, thermomechanical, x-ray diffraction, and NMR studies. Poly[bis(trifluoroethoxy)-phosphazene], $[(CF_3CH_2O)_2PN]_n$ (I), is taken as an example of a typical alkoxy phosphazene and poly[bis(p-chlorophenoxy)phos-phazene], $[(p-ClC_6H_4O)_2PN]_n$ (II) as an example of an aryloxy phosphazene.

II. EXPERIMENTAL

The samples used in this study were synthesized by the two step procedure which involves the preparation of polydi-chlorophosphazene by thermal polymerization of cyclic trimer followed by nucleophilic substitution with a selected sodium alkoxide or aryloxide. Detailed procedures have been given elsewhere (1, 2). All samples were submitted for elemental analysis to verify that substitution was essentially complete. Calorimetric measurements were carried out using the Perkin-Elmer DSC II. Thermomechanical analyses in the penetrometer and thermal expansion modes were run with the Perkin-Elmer TMS1 assembly. X-ray patterns were recorded with a Warhus flat-film camera with copper radiation using a heated sample holder. Solid state NMR measurements were made at 20 MHz using a spectrometer which has been described elsewhere (4).

III. RESULTS AND DISCUSSION

A. Initial Observations

Evidence for the appearance of two first order transitions in several polyphosphazenes was first reported in Allen et al. (5) based on observations carried out by optical microscopy and DTA. The various aspects of the behavior (6) are best observed with I. Solution cast films of I show large scale spherulitic morphology which remains unaltered on heating through $T(1) = 90°C$ up to the melting temperature $T_M = 240°C$. Recrystallization after heating above T_M produces needle-shaped crystals rather than the original spherulitic morphology. Allen showed that the crystalline x-ray diffraction pattern of I on heating above 90°C is reduced to a single sharp equatorial line at 11Å and diffuse meridional scattering at 4-5Å, indicating the transition to a mesomorphic state with lateral order and longitudinal disorder.

Dynamic mechanical measurements reported by Allen et al. (5) show that a drastic reduction in modulus occurs at $T(1)$. In fact the samples can be molded near or above $T(1)$ to form coherent films. Corresponding changes have been observed in the structure and properties of most other crystalline polyphosphazenes.

B. Calorimetric and Thermal-Analytical Measurements

Further details about the transition behavior can be obtained from the results of calorimetric and thermal-analytical methods (6). It will be convenient to refer to the behavior at T(1) as melting, taking care to distinguish the transition to the liquid state as true melting where confusion might arise. As shown in Figure 1, the melting endotherm at T(1) for I is initially broad but sharpens appreciably with repeated cycling to a temperature just above T(1) or with annealing above T(1), accompanied by a gradual increase in T(1). When the sample is recrystallized by slow cooling from the true melt (run 3), there is a dramatic increase in the area and sharpness of the endotherm at T(1) with a more than 10°C increase in transition temperature.

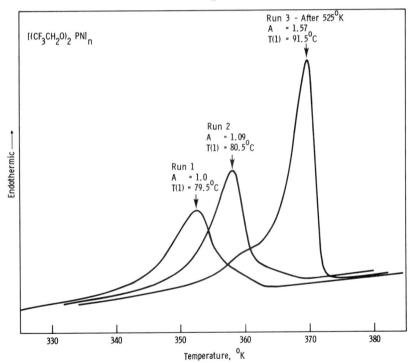

Fig. 1. Influence of thermal cycling on the crystalline to mesomorphic transition in sample I. Successive curves displaced 5° for clarity.

Other work has shown that if this sample is now repeatedly cycled through T(1), a progressive broadening of the endotherm and reduction in T(1) occurs. The original features of the endotherm can be restored by recrystallization from the melt, proving that the changes are not due to degradation. These observations indicate that the organization of the mesomorphic state can be modified by annealing or by the disruptively large volume change occurring at T(1) and, in turn, that the mesomorphic state exerts a strong influence on the crystalline structure.

A comparison of the endotherm at T(1) and T_M is shown in Figure 2. The enthalpy change at T(1) is ten fold greater than at T_M, and the peak is appreciable sharper at the lower transition. The recrystallization processes at T(1) and T_M show strong supercooling effects similar to those observed with crystalline polymers on cooling from the melt (6).

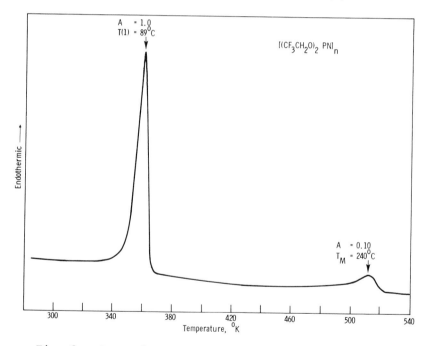

Fig. 2. *Comparison of endotherms at T(1) and T_M for sample I after recrystallization from the melt.*

The endotherm at T(1) for II shows a similar response to repeated cycling above T(1) = 160°C and to high temperature annealing. Allen et al. reported T_M = 350-400°C from optical microscopy and DTA. A careful search for T_M was made by DSC to 410°C but failed to show any evidence of a melting peak. This was taken as an indication that the enthalpy change at T_M for II is close to zero.

The TMA system with thermal expansion probe was used to estimate the volume changes at T(1) and T_M (6). The sample was cut to fit entirely under the probe and the probe was seated by thermal cycling with a weight 1 to 5g above the

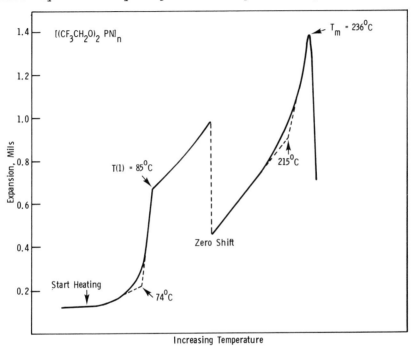

Fig. 3. Thermal expansion curve obtained by TMA for sample I; thickness, 12.5 mil

Table 1.
COMPARISON OF THERMAL TRANSITION DATA

Sample	T(1) °C	T_M °C	$\Delta H(T(1))$ cal/g	$\Delta H(T_M)$ cal/g	$\Delta V(T(1))$ %	$\Delta V(T_M)$ %
Poly[bis(trifluoro-ethoxy)phosphazene] (Ref. 6)	92	240	8.6	0.8	5	6
Poly[bis(p-chloro-phenoxy)phosphazene] (Ref. 6)	169	356	6.6	~0	3.5	5.7
Poly(diethylsiloxane) (Ref. 7)	-5	20	2.4	0.36		
Teflon (Ref. 8)	19-30	327	8.5	13.7	1	20

neutral buoyancy point. As shown in Figure 3 for I, the expansion at T(1) is well defined and almost the full course of the expansion at T_M, starting from below 200°C, can be observed before the sample softens and flows. With care similar results can be obtained for II. In contrast to the marked disparity in the heats of transition at T(1) and T_M, the volume changes were of the order of 4 to 6% at the two transitions. The results of the thermal analyses are summarized in Table 1 where comparisons are made with several other polymers which exhibit a transition to a mesomorphic state.

C. Transition Temperatures

Table 2 lists the transition and the decomposition temperatures for a selected group of polyphosphazenes (3, 9).

Table 2.

THERMAL ANALYSIS DATA FOR VARIOUS POLYPHOSPHAZENES

$$\begin{array}{c} OR \\ | \\ +P = N\}_n \\ | \\ OR \end{array}$$

RO-	T_g (°C)	T(1) (°C)	T_M (°C)	T_D (°C)
CH_3CH_2O-	-84			
CF_3CH_2O-	-66	92	240	360
C_6H_4O-	6	160	390	380
$m-ClC_6H_4O-$	-24	90	370	380
$p-ClC_6H_4O-$	4	169	365	410
$m-CH_3C_6H_4O-$	-25	90	348	350
$p-CH_3C_6H_4O-$	0	152	340	310
$p-CH_3CH_2C_6H_4O-$	-18	43	*	285
$p-(CH_3)_3C \; C_6H_4O-$	48	237	345	350
$3,4-[CH_3]_2C_6H_3O-$	-5	96	325	315
$3,5-[CH_3]_2C_6H_3O-$	6	67	320	340
$p-C_6H_5CH_2C_6H_4O-$	-3	109		320

*Decomposes with expansion

One important conclusion which can be drawn from this table and more complete surveys of the data presented elsewhere (9), is that almost all the crystalline poly(organophosphazenes) show a T(1) transition indicating that this transition is not limited to certain alkoxy or aryloxy derivatives. The values of T_M determined by penetrometer analysis, in many cases, occur near the decomposition temperature T_D. This implies that T_M

can only be taken as an indication that structure capable of
supporting the weighted probe persists to this temperature.
Several generalizations regarding the dependence of T(l) on
composition can readily be drawn from the results in Table 2.
The effect of meta compared to para substitution in the aryloxy
polymers on lowering T_g and T(l) is particularly striking, as
illustrated by several examples in the table. Note, however,
that T_M is virtually insensitive to these positional substitution differences on the phenyl ring.

D. X-Ray Results

The first direct evidence regarding the structure of the
mesomorphic state is the x-ray pattern at $90°C$ reported by
Allen et al. (5) for I as discussed earlier. More recent work
(10) on II and the related poly[bis(m-chlorophenoxy)phosphazene]
(III) revealed two sharp equatorial lines in each of these
polymers between T(l) and T_M. The structure is determined to
be pseudohexagonal from the fact that these two lines are near
the ratio 1.732 expected for d_{100}/d_{110} in a metrically
hexagonal cell. In addition to the strong 100 and weaker 110
lines a third sharp but faint line may be indexed at 210
strengthening the assignment. Diffuse meridional scattering
at 4.1 to 4.5A is attributed to hkl reflections broadened by
rotational or longitudinal disorder.

To understand the mechanism of the transformation from
the crystalline to mesomorphic state a knowledge of the
crystalline structure is required. Unit cell assignments have
been reported for several poly(organophosphazenes) (11) but
the only structural analysis is that proposed by Bishop and
Hall (12) for II. However, their results must be viewed with
some reservations due to the rather large residual differences
between calculated and observed intensities. The polymer
chain is assumed to occupy a planar cis-trans conformation
consistent with a repeat distance of about 4.9A for two monomer
units which appears in virtually all crystalline poly(organophosphazenes). One consequence of the cis-trans conformation
is that all bonds from phosphorus to ligands occur at the same
angle to the chain axis, thus, conferring a sense of direction
on the chain. The unit cell is orthorhombic with one chain at
the center and one at each corner. The direction of the central
chain is opposite that of the four surrounding chains.

The transformation geometry (10) of II from the orthorhombic crystal to the mesomorphic state is indicated in Figure 4.
The a/b ratio increases from 1.55 to about 1.73 in the
hexagonal state due to a 9% expansion in the a and 20% expansion in the b directions of the Bishop and Hall orthorhombic

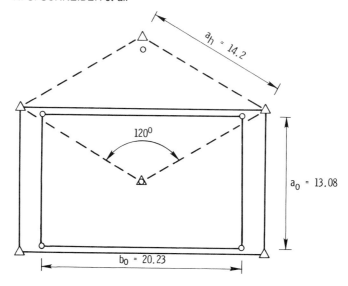

o Chain Sites, Orthorhombic Crystal

△ Chain Sites, Pseudohexagonal Phase

TRANSFORMATION FROM ORTHORHOMBIC
TO PSEUDOHEXAGONAL STRUCTURE

Fig. 4.   Geometry for the transformation of the orthorhombic crystal to the hexagonal mesomorphic structure for sample II.

lattice. As a result the diagonals of the unit cell make an angle of 120° with each other and the structure is metrically hexagonal. Although the term pseudohexagonal was used to describe this structure in earlier work (10), since three-fold symmetry had not been established, work to be discussed shortly indicates that the mesomorphic state is symmetrically as well as metrically hexagonal. Examination of the x-ray data for several other poly(organophosphazenes) strongly suggests that the hexagonal structure is a general characteristic of the mesomorphic state in these polymers and that the single diffraction line usually observed can be identified as the strong 100 reflection which appears in II and III. Correspondingly, the values of the interchain spacing a_h calculated from the 100 spacings indicate that the hexagonal packing is controlled by side group size (13). Polymer I with a relatively small side group has a value $a_h = 11.7$–11.8Å; II, III and $(3,4(CH_3)_2C_6H_3O)_2PN]_n$ have $a_h = 14.0$–14.1Å; while $[(\underline{p}\text{-}C_6H_5CH_2C_6H_4O)_2PN]_n$ with a very bulky side group has the much larger value $a_h = 19.4$Å.

E. NMR Studies

The x-ray data, in itself, cannot distinguish between
static positional disorder and disorder involving rapid side
chain and back bone motions. However, in analogy with
transitions in such polymers as poly(tetrafluoroethylene) (14)
and trans-1,4-poly(butadiene) (15), it seemed reasonable to
postulate that the T(1) transition in poly(organophosphazenes)
involves the onset of rapid rotational motions. To test this
hypothesis, poly[bis(trifluoroethoxy)phosphazene] (I) was
selected for examination by wide line NMR. This polymer
contains three nuclei (^{1}H, ^{19}F, ^{31}P) which can be readily
studied by NMR. The resonance of the first two nuclei are
sensitive to the side group motions, while the ^{31}P resonance
provides information about backbone motions, making this a
particularly advantageous system. The results reported here,
although preliminary, definitely indicate that the T(1)
transition is accompanied by rapid re-orientational motion.
A more complete account of this work is to be published (16).

The polymer sample was dissolved in tetrahydrofuran,
precipitated with methanol, and compacted by centrifugation.
NMR derivative spectra were obtained in continuous wave mode
at 20 HMz for each of the three nuclei at 20° to 120°. The
experimental line widths are listed in Table 3 . For both
side chain nuclei (^{19}F and ^{1}H), the spectra at 20° consist of

Table 3

NMR LINE WIDTHS (OERSTEDS)

| Nucleus | $[(CF_3CH_2O)_2PN]_n$ | |
	T = 20°C	T = 120°C
^{19}F	2.3,0.7	0.4
^{1}H	6,0.8	0.6
^{31}P	2.4	1.1

two lines, one broad and one narrow, having a common center.
Pulse techniques were used to establish that the two lines
have their origin in separate portions of the sample. The
broad line probably arises from crystalline material while
the narrow line originates in amorphous material. Upon
raising the temperature to 120° (well above T(1) at 92°), the
broad line disappears for both the ^{19}F and ^{1}H side chain
spectra, leaving single lines somewhat narrower than the narrow
components at 20°. It is conceivable that the narrow line

consists of two components which are not resolvable due to the magnetic field inhomogeneity of about 0.1 Oe. The observed line narrowing indicates that the transition at T(1) increases the mobility of both the crystalline and amorphous regions. Furthermore, the side chain lines at 120° are narrower than one would predict from side group motions alone, indicating that the backbone motions are also involved.

The ^{31}P absorption shows no evidence of separate crystalline and amorphous components, either at 20° or 120°. There is an experimental problem, however, in that the ^{31}P signal is easily saturated, so that lines broader than those reported would be lost in the noise. Thus, we presume that the crystalline phase signal is lost at 20°. The narrowing between 20° and 120° results from increased rotational motion, but the line is still broader than those of the side chain nuclei. This is because the motions of the backbone atoms do not randomize their mutual orientations, so that interactions between backbone atoms contribute to the phosphorus line broadening. The results are consistent with a simplified model presuming rapid rotation about the backbone axes, which will be reported in greater detail (16).

IV. CONCLUSIONS

The properties of the crystalline polyphosphazenes are dominated by the transition from the crystalline to the mesomorphic state at T(1). The softening of the polymer at this temperature sets the upper limit to useful engineering properties and represents the temperature at which the material may be formed as a thermoplastic, but by shear of the semi-rigid structure, rather than by melt flow. X-ray diffraction results and preliminary solid state NMR data has shown that the mesomorphic state involves rapidly rotating chains in a hexagonal lattice, exhibiting lateral order but longitudinal disorder. Thus, the mesomorphic state in the polyphosphazenes may be identified with the behavior of a small group of varied polymers which also exhibit a pseudo-hexagonal state arising from dynamic disorder. Much remains to be done experimentally in defining the structure, extent of motion, the thermodynamics and kinetics of the transitions at T(1) and T_M. This information should lead to a sound understanding of the characteristic behavior of the polyphosphazenes, particularly with respect to the differences from other polymers which exhibit related transition phenomena.

V   REFERENCES

1. Allcock, H. R., Kugel, R., and Valan, K. J., Inorg. Chem. 5, 1709 (1966).
2. Singler, R. E., Hagnauer, G. L., Schneider, N. S., LaLiberte, B. R., Sacher, R. E., and Matton, R. W., J. Polym. Sci., Polym. Phys. Ed. 12, 433 (1974).
3. Singler, R. E., Schneider, N. S., and Hagnauer, G. L., Polym. Eng. Sci. 15, 321 (1975).
4. Lessler, A. J., Alexander, M. N., Sagalyn, P. L., and Walker, N., J. Chem. Phys. 63, 3971 (1975).
5. Allen, G., Lewis, C. J., and Todd, S. M., Polymer 11, 44 (1970).
6. Schneider, N. S., Desper, C. R., and Singler, R. E., J. Appl. Polym. Sci. 20, 3087 (1976).
7. Beatty, C. L., and Karasz, F. E., J. Polym. Sci., Polym. Phys. Ed. 13, 971 (1975); Beatty, C. L., Pochan, J. M., Froix, M. F., and Hinman, D. D., Macromolecules 8, 547 (1975).
8. Furukawa, G. T., Mc Coskey, R. E., and King. G. J., J. Res. Nat. Bur. Stand., Sect. A. 49, 273 (1952); Starkweather, J. W., and Boyd, R. H., J. Phys. Chem. 64, 410 (1960); Mc Cane, D. I., in "Encyclopedia of Polymer Science and Technology" (H. F. Mark, N. G. Gaylord and N. N. Bikales, Eds.), Vol 13, pp. 628, Wiley-Interscience, New York, 1970.
9. Schneider, N. S., Desper, C. R., and Beres, J. J., in "Liquid Crystalline Order in Polymers" (A. Blumstein, Ed.) Academic Press, New York, in press.
10. Desper, C. R., and Schneider, N. S., Macromolecules 9, 424 (1976)
11. Stroh, Jr. E. G., Ph.D. thesis, The Pennsylvania State Univ., (1972).
12. Bishop, S. M., and Hall, I. H., Br. Polym. J. 6, 193 (1974).
13. Desper, C. R., Schneider, N. S., and Higginbotham, E., J. Poly. Sci., Polymer Letters Edn. in press.
14. Hyndman, D., and Origlio, G. F., J. Appl. Phys. 31, 1849 (1960); McCall, D. W., Douglass, D. C., and Falcone, D. R., J. Phys. Chem. 71, 998 (1967).
15. Iwayanagi, S., and Miura, I., Rept. Progr. Polymer Phys. Japan 8, 303 (1965).
16. Alexander, M. N., Desper, C. R., Sagalyn, P. L., and Schneider, N. S., Macromolecules, in press.

POLY(ORGANOPHOSPHAZENES) DESIGNED FOR BIOMEDICAL USES

Harry R. Allcock

The Pennsylvania State University

University Park, Pennsylvania 16802

Abstract

Three classes of poly(organophosphazenes) have been synthesized with biomedical uses in view. These include hydrophobic poly(fluoroalkoxyphosphazenes), such as $[NP(OCH_2CF_3)_2]_n$, poly(amino acid ester phosphazenes), such as $[NP(NHCH_2COOC_2H_5)_2]_n$, and coordination complexes of the type $[NP(NHCH_3)_2]_x(PtCl_2)_y$. The platinum-containing species were designed as water-soluble polymeric anticancer agents, and the amino acid ester substituted polyphosphazenes were designed to degrade hydrolytically to amino acid, phosphate, alcohol, and ammonia. The structure and reactions of the polymers are discussed.

Introduction

Severe limits exist to the number of organic-type polymers that can be used for biomedical purposes. These limits are imposed by detrimental chemical interactions that take place between some polymers and blood or tissues, by unsatisfactory surface physical properties, and by the scarcity of organic polymers that are either soluble in water to give biocompatible solutions or degradable in a living system to harmless products.

Poly(organophosphazenes) (III-V) are almost unique with respect to the range of different polymers that can be readily synthesized from one backbone structure. Moreover, the method of synthesis permits different polymers with specific properties to be "tailor-made" by changes in the side group structure. Since our initial discovery of the first synthesis routes to poly(organophosphazenes) (1-3), considerable research and development work has been directed to the synthesis of mixed-substituent poly(organophosphazene) elastomers for technological applications (4-7). Much of our recent work has involved the synthesis of poly(organophosphazenes) that may be useful as (a) biologically inert materials for biomedical reconstruction or artificial organ usage; (b) polymers that will degrade slowly in the body to harmless, readily excretable products; and (c) water-soluble high polymers that can function as carriers for drug molecules. This paper will provide a brief review of this work.

General Synthesis Route

The general synthetic method involves the thermal polymerization of hexachlorocyclotriphosphazene (I) to high
molecular weight poly(dichlorophosphazene) (II), followed by
replacement of the halogen in II by treatment with nucleo-
philes, such as alkoxides, aryloxides, or amines (1-3).   In
view of the broad range of nucleophilic reagents that can
participate in this reaction, the synthetic versatility of
the method is obvious.

Biologically Acceptible Polyphosphazenes

An ideal material for the construction of artificial
heart pumps, heart valves, replacement blood vessels, etc.,
should be flexible or elastomeric and should have a minimal
tendency to initiate the clotting of whole blood or lymph.
Preliminary in vitro screening tests performed on a variety of
phosphazene polymers with the use of Lindholm cells (8,9)
indicated that phosphazene high polymers that contained
fluoroalkoxy substituent groups, for example, $[NP(OCH_2CF_3)_2]_n$,
were particularly promising.   One advantage foreseen for
these polymers over silicones, is the lower tendency of the
phosphazenes to absorb lipids.

A number of polymers have been synthesized, of general formula IV, in which amino acid ester units are present as substituent groups (10).   Glycine ethyl ester, leucine methyl ester, alanine methyl ester, and phenylalanine methyl ester groups have been introduced, both as homopolymers and as mixed substituent polymers containing methylamine groups (VI-X).   The compound studied in the greatest detail is shown as VI.

$$
\left[\begin{array}{c} NHCH_2COOC_2H_5 \\ | \\ -N=P- \\ | \\ NHCH_2COOC_2H_5 \end{array} \right]_n
\qquad
\left[\begin{array}{cccc} NHCH_2COOC_2H_5 & & NHCH_3 \\ | & & | \\ -N=P\!-\!-\!-\!-\!-\!N=P- \\ | & & | \\ NHCH_3 & & NHCH_3 \end{array} \right]_n
$$

VI   VII

$$
\left[\begin{array}{c} CH_3 \\ | \\ NHCHCOOCH_3 \\ | \\ -N=P- \\ | \\ NHCH_3 \end{array} \right]_n
\qquad
\left[\begin{array}{cccc} CH_2CH(CH_3)_2 & & \\ | & & NHCH_3 \\ NHCHCOOCH_3 & & | \\ | & & \\ -N=P\!-\!-\!-\!-\!-\!N=P- \\ | & & | \\ NHCH_3 & & NHCH_3 \end{array} \right]_n
$$

VIII   IX

$$
\left[\begin{array}{c} CH_2C_6H_5 \\ | \\ NHCHCOOCH_3 \\ | \\ -N=P- \\ | \\ NHCH_3 \end{array} \right]_n
\qquad
\left[\begin{array}{c} NHCH_3 \\ | \\ -N=P- \\ | \\ NHCH_3 \end{array} \right]_n
$$

X   XI

Polymer VI is a leathery material at room temperature.   The glass transition temperature was $-23°C$ and the $\overline{M}_n$ values varied between 2×10^6 and 5×10^6 for different samples.

Hydrolysis at pH 7.5 at temperatures of 25°C-50°C resulted in
a slow liberation of free glycine and a degradation of the
chain.   The ultimate hydrolysis products are believed to be
amino acid, ethanol,phosphate, and ammonia.   However, slow
chain cleavage also occurs by a second mechanism that involves
attack on the backbone by COOH groups generated by surface
hydrolysis.   This leads to a slow spontaneous molecular
weight decline at 25°C or higher temperatures.   The total
replacement of chlorine in II by leucine ester, alanine ester,
or phenylalanine ester groups proved to be more difficult
because of steric hindrance effects.   However, the remaining
P-Cl bonds could be replaced by methylamino residues.   These
polymers strongly bind acids, such as hydrogen chloride,
presumably by protonation of the skeletal nitrogen atoms.

Phosphazenes as Carrier Molecules for Drugs

Poly[bis(methylamino)phosphazene] (XI) is a water-
soluble polymer that bears coordination sites on both the
side group and chain nitrogen atoms (11,12).   Hence, this
material provides an unusual substrate for the coordination
binding of metal atoms.   Polymer XI was allowed to react
with K_2PtCl_4 and 18-crown-6-ether in organic media to yield a
coordination complex of structure XII (11,13).

XII   XIII

The binding of the platinum to the skeletal nitrogen atoms of
the polymer was inferred from the structure of a cyclic
tetrameric analogue (XIII).   An X-ray single crystal study
of XIII showed unambiguously that the platinum was bound in a
transannular manner to two 2,6-skeletal nitrogen atoms (14).
The spectroscopic similarities between XII and XIII strongly
suggested a similar mode of binding in XIII.   Compounds XII
and XIII showed tumor inhibitory activity in initial anticancer screening tests against mouse P388 lymphocytic leukemia

and in the Ehrlich Ascites tumor regression test (15).   The
reaction of aminophosphazenes, such as XIII with K_2PtCl_4
yields tetrachloroplatinate salts of the phosphazenes (13).

Conclusions

The synthetic versatility and technological usefulness
of high molecular weight polyphosphazenes is fully established.
The biomedical utility of these polymers is only beginning to
be investigated.   The preliminary exploration of ways in
which polyphosphazenes might be used in medicine has indicated
that eventually these polymers may prove more useful than
polysiloxanes because of their greater chemical adaptability.
However, the use of any new polymer system in biomedicine is
dependent on the results of long term, in vivo tests.   The
prospects seem encouraging, but the direction of future
developments will be determined by the more controlled
biological programs now being initiated.

Acknowledgements

Financial support of this work by the National Heart
and Lung Institute of NIH and by the Army Research Office,
Durham, is gratefully acknowledged.

References

1. H. R. Allcock and R. L. Kugel, J. Amer. Chem. Soc., 87,
    4216 (1965).
2. H. R. Allcock, R. L. Kugel, and K. J. Valan, Inorg. Chem.,
    5, 1709 (1966).
3. H. R. Allcock and R. L. Kugel, Inorg. Chem., 5, 1716 (1966).
4. S. H. Rose, J. Polym. Sci., B6, 837 (1968).
5. D. P. Tate, J. Polym. Sci., Polym. Symp., 48, 33 (1974).
6. R. E. Singler, N. S. Schneider, and G. L. Hagnauer, Polym.
    Engineering Sci., 15, 321 (1975).
7. H. R. Allcock, Chem. Rev., 72, 315 (1972); "Phosphorus-
    Nitrogen Compounds", Academic Press, New York, 1972;
    Science, 193, 1214 (1976).
8. R. G. Mason, Biomat., Med. Dev., Art. Org., 1, 131 (1973).
9. H. R. Allcock, K. M. Smeltz, and S. D. Wright, unpublished
    results.
10. H. R. Allcock, T. J. Fuller, D. P. Mack, K. Matsumura, and
    K. M. Smeltz, Macromolecules, in press (1977).

11. H. R. Allcock, R. W. Allen, and J. P. O'Brien, J. Chem.
    Soc., Chem. Comm., 717 (1976).
12. H. R. Allcock, W. J. Cook, and D. P. Mack, Inorg. Chem.,
    11, 2584 (1972).
13. H. R. Allcock, R. W. Allen, and J. P. O'Brien, J. Am. Chem.
    Soc., in press (1977).
14. R. W. Allen, J. P. O'Brien, and H. R. Allcock, J. Am. Chem.
    Soc., in press (1977).
15. Tests performed by Dr. Iris H. Hall, Div. of Medicinal
    Chemistry, Univ. of North Carolina.

BIOCOMPATIBILITY OF EIGHT POLY(ORGANOPHOSPHAZENES)[1,2]

C.W.R. Wade, S. Gourlay, R. Rice and A. Hegyeli
US Army Medical Bioengineering
Research & Development Laboratory
R. Singler and J. White
Army Material and Mechanics Research Center

Eight poly(organophosphazenes) $(R_2P = N)_x$ with widely different physical and mechanical properties were cast into films and implanted intramuscularly into rats for periods of 7 and 28 days. Silicone sections of the same shapes and sizes were similarly implanted as controls. The tissue implant sites were excised and examined macroscopically and microscopically. Each implant site was characterized for cell types, total cellular density and thickness of cellular reaction. The examinations revealed no signs of adverse tissue reaction. Correlation of the data, according to the method of Sewell, Willand and Craver, showed a remarkably similar tissue response to each of the phosphazenes, irrespective of the organo substituents, and to the silicone controls. For each material, the biological toxicity was classified as slight.

Because the tissue reaction induced in rats by these poly(organophosphazenes) was no stronger than that of SILASTIC[R], a most widely used biomaterial, this new group of polymers merits further studies to assess their potential biomedical applications.

I. INTRODUCTION

The increasing use of synthetic materials for the surgical repair, replacement or restoration of the function of diseased or damaged tissue is a persistent stimulant for research in the development of new materials. To satisfy rigorous medical requirements, the selected materials ideally must be compatible with the biological milieu and mechanical

---

[1]In conducting the research described in this report, the investigators adhered to the "Guide for Laboratory Animal Facilities and Care", as promulgated by the Committee on the Guide for Laboratory Animal Facilities and Care of the Institute of Laboratory Animal Resources, National Academy of Sciences - National Research Council.

[2]The opinions or assertions contained herein are the private views of the authors and are not to be construed as official or as reflecting the views of the Department of the Army or the Department of Defense.

components and should show zero or controlled biodegrada-
tion, depending upon the application. The materials must be
interfaceable with the surrounding and connecting tissues
with no more than minimal adverse reactions. Polymers, ce-
ramics and metals are currently used in such medical applications but no class of materials is without serious problems. Consequently, the search for better materials
continues.

Poly(organophosphazenes), substances of the class
$[R_2P = N]_x$, have attracted increasing attention in recent
years as stable polymers for fundamental chemical study[1] and
as materials with a host of potential applications[2-8].
Variations in mechanical properties can be achieved simply
by exchanging R_2- groups during synthesis.

Although many publications and reviews[9-10] have
appeared concerning the preparation,[11-15] characterization[16]
and utilization of these interesting linear polymers, no definitive studies have been reported on their potential as
stable biomedical materials[17]. Nor has there appeared an
account of in vivo evaluation of these materials as a step
toward assessing their potential use in medical applications.
The present communication reports on a comparative study of
the biocompatibility of the widely used silicone rubber,
SILASTIC[R] (Dow Corning's Medical Grade Silicone) and eight
stable poly(organophosphazenes)[5].

II. EXPERIMENTAL

A. Poly(organophosphazene) Preparation

The eight polymers (Table 1) were prepared and characterized at the Army Material and Mechanics Research Center
according to published methods[10,11,15]. Substituent groups
and selected properties are also included in the table.

TABLE 1

$$\left[\begin{array}{c} R \\ -P = N- \\ R \end{array} \right]_x$$

SOME PHYSICAL PROPERTIES OF THE POLYPHOSPHAZENES

POLYMER NO.	R	[n] dl/g[THF]	Tg,°C	Td,°C	SOLVENTS
1	CF$_3$CH$_2$O-	1.3	-66	350	THF, Acetone
2	m-MePhO-	2.1	-25	340	THF, CHCl$_3$
3	p-sec-BuPhO-	2.9	-	-	THF
4	p-t-BuPhO-	1.0	50	352	THF
5	PhNH-	1.4	105	-	THF, Benzene, CHCl$_3$
6	p-MeOPhNH-	1.3	92	266	THF, Benzene, CHCl$_3$
7	p-n-BuPhNH-	0.9	53	253	THF, Benzene, CHCl$_3$
8	m-ClPhNH-	1.3	80	253	THF, Benzene, Acetone

[n]: Inherent Viscosity   Tg: Glass Transition   Td: Decomposition

THF: Tetrahydrofuran

B. Film Preparations

A sample of each of the poly(organophosphazenes) was dissolved in tetrahydrofuran, THF, to give a 10% solution. The solutions were filtered, cast onto sheets of poly(tetrafluoroethylene) to give films 5 to 10 mils thick. After air-drying, the films were dried in vacuo at 40°C and 0.5 mm of Hg for 24 hours. The dried films were cut into implants 5 x 5 mm^2 and rinsed in 50% aqueous ethyl alcohol followed by repeated rinsings with distilled water. Control samples of 5-mil thick SILASTIC[R] silicone rubber sheeting were similarly cut into implant pieces, washed with mild soap and rinsed with distilled water to remove the starch on the silicone surface. Following air-drying, the test samples and controls were packaged and sterilized with 1.5 Mrad of ^{60}Co gamma radiation.

C. Implantation in Animals

Thirty-nine male Sprague-Dawley derived rats weighing between 200 and 250 g were anesthetized with 0.5 cc/Kg Innovar-Vet[R] (Pitman-Moore) and their dorsal areas were clipped and prepared with Betadine[R] Solution (Pudue-Frederick). A midline incision was made on each rat and the skin and panniculus were reflected from both gluteal areas. Incisions were made in the direction of the fibers in the gluteus maximus muscles on either side and into the fascial planes between these muscles. The underlying gluteus medius muscles were separated bluntly. Implants rinsed in normal saline were inserted atraumatically into these spaces (Fig. 1). A poly(organophosphazene) test specimen was implanted on the right side in each rat and a silicone rubber control was implanted on the left side. Four rats were implanted with sample number 4 (See Table 1), and 5 rats each were implanted with the other test samples. The skin wounds were closed with 9-mm stainless steel clips. All rats were housed individually and allowed food and water ad libitum.

Two rats were sacrificed at 7 days and 3 rats at 28 days (2 rats at 28 days for number 4) for each of the poly(organophosphazenes). Each gluteal muscle group containing an implant was removed en bloc and fixed in buffered 10% formalin. Specimens were divided into 2 groups, and each group was processed and sectioned (8 μm) routinely. Slides were prepared and one group was stained with hematoxylin/eosin and the other with Masson's trichrome. All sections were examined microscopically.

Fig. 1. Polymer film implantation into the rat showing the dorsal incision with reflected skin and panniculus and the incision parallel to the muscle fibers in the right gluteus maximus muscle. The circular inset shows a 5 x 5 mm^2 test sample being inserted atraumatically into the intermuscular fascial plane beneath the gluteus maximus and the gluteus medius muscles.

D. Grading of Tissue Response

The tissue response was graded according to a modified system of Sewell et al[18]. In our grading system an 0.32-mm diameter circular field was used as the reference standard instead of the 430x "high power field" used by Sewell so that two microscopes could be used concurrently. The scoring for capsular thickness, cell type and cell density were the same as recommended by Sewell. In each case the tissue reaction was measured adjacent to the undersurface of the implant and next to the uninjured gluteus medius muscle. The scores were summed and the statistical significance of resulting values for the total tissue reactions were determined using the Student's t test[19].

III. RESULTS

A. Physical and Analytical Properties

Some of the physical characteristics of the eight poly(organophosphazenes) are listed in Table 1. Glass transition temperatures range from $-66°C$ to $105°C$. The highest glass transition temperatures are observed for anilino and substituted anilino derivatives, all of which are brittle polymers, and the lowest glass transition is for the elastomeric fluoroethoxy polymer. The polymers are thermally stable up to $250°C$, above which degradation begins to occur. All eight polymers are soluble in tetrahydrofuran, anilino derivatives are soluble in benzene, and some phenoxy and anilino derivatives are soluble in chloroform. Viscosities of solutions of the polymers varied from 0.9 to 2.9 dl/g, due mostly to variation in the molecular weight of the starting polymer, poly(dichlorophosphazene), and to the substituents. In the infrared, there is a distinct shift in frequency for P=N vibrations, depending upon whether the polymer has anilino, fluoroethoxy or phenoxy substituents.

Data are given for the elemental analyses of these compounds in Table 2. It is readily obvious that complete replacement of Cl groups did not occur, and using the value

TABLE 2

Elemental Analyses of the Poly(phosphazenes)

POLYMER NO.	CALCULATED (%)				FOUND (%)			
	C	H	Cl	N	C	H	Cl	N
1	19.77	1.66	0	-	19.75	1.48	0.03	-
2	64.86	5.44	0	-	64.67	5.44	0.91	-
3	69.95	7.63	0	-	70.39	7.44	0.50	-
4	69.95	7.63	0	-	69.48	7.44	0.11	-
5	-	-	0	-	-	-	0.76	-
6	58.12	5.59	0	14.53	57.18	5.69	0.15	14.31
7	70.34	8.28	0	12.31	68.97	8.18	0.32	12.29
8	48.34	3.39	23.79	14.10	45.69	3.34	23.30	14.09

of 0.9% chlorine (polymer No. 2) indicates one Cl atom
(35.5 g) per 3900 g of the aryloxy-polymer. The value
"23.30 found" (23.79 calculated) is a result of the Cl group
on the benzene ring and not of replaced Cl attached to the
polymer backbone. Characteristic properties of SILASTIC[R]
are documented in the literature[9] and were not determined.

B. Biomedical Evaluations

There were no deaths of rats during the course of the
experiments and all the rats appeared to be healthy at the
time of sacrifice. All implants were encapsulated by thin
fibrous tissue, Figure 2, and differences in appearance
between test and control sites at both 7 and 28 days were
minimal. Figures 2 and 3 are low and high magnification
photomicrographs showing typical tissue reactions to
poly(organophosphazene) and SILASTIC[R]. Examination of the
implant sites showed little difference in tissue reaction
between the test materials and the controls at either time.
The reaction was uniformly that of a thin fibrous capsule
with few mononuclear phagocytes (more at 7 than at 28 days).
Foreign body giant cells were rarely present, and no neutro-
phils, eosinophils or lymphocytes were found at the implant
site. No changes in the microscopic structure of the gluteus
medius muscle were observable. Implants were sometimes lost
from the capsule during preparation of the specimens for
microscopic examination. They frequently separated during
sectioning, staining and cover glass mounting, and several
of the poly(organophosphazenes) were dissolved by the solvents used in those processes. Tissue grading scores obtained by correlating the cell types, thickness of cellular
film, and cell density, are listed in Table 3. The mean
scores for the individual poly(organophosphazenes) and for
their controls fall in most cases into toxicity grade 2 of
Sewell et al (slightly toxic, 16-32) points[18]. The variance
was less among the poly(organophosphazenes) than among the
SILASTIC[R] controls. The total mean[10] tissue response scores
for all 8 poly(organophosphazenes) were 26.0 ± 4.1 S.D. and
27.9 ± 2.4 S.D. at 7 and 28 days, respectively. The corresponding values for the SILASTIC[R] controls were 29.0 ± 9.9
S.D. and 27.9 ± 3.8 S.D. Paired t testing showed no significant difference in the tissue responses to the test and
control materials at either 7 or 28 days (t = 1.375, 15 d.f.
at 7 days; t = 0.080, 22 d.f. at 28 days). Likewise, unpaired t testing for differences in the tissue response between the 7 and 28-day tests showed no significant differences for either the test materials or the controls
(t = 1.924, 37 d.f. for the poly(organophosphazenes);
t = 0.499, 37 d.f. for SILASTIC[R].

TABLE 3

Tissue Reaction in Rats to Implanted Films

of Polyphosphazenes and Silastic[R] *

Sample Number (See Table I)	7 Days Polyphosphazene	Silastic[R]	28 Days Polyphosphazene	Silastic[R]
1	32.2	30.2	25.7	25.9
2	25.0	22.5	30.6	30.7
3	27.9	42.0	28.3	27.1
4	24.5	28.0	26.0	22.0
5	21.5	26.0	28.3	28.7
6	29.5	36.2	26.7	26.3
7	24.5	25.5	26.7	26.3
8	22.5	21.5	27.0	26.0
Mean	26.0	29.0	27.9	27.9
S.D.	4.1	9.9	2.4	3.8
N	16	16	23	23

IV. DISCUSSION

The phosphazene polymers used in the study, namely:
(1) Bis(2,2,2-trifluoroethoxy), (2) bis(3-methylphenoxy),
(3) bis(4-sec-butylphenoxy), (4) bis(4-t-butylphenoxy), (5)
bis(phenylamino), (6) bis(4-methoxyphenylamino), (7) bis(4-
n-butylphenylamino), and (8) bis(3-chlorophenylamino), like
the silicones, are inorganic materials to which organo substituents have been chemically added. As with the silicones, these organo groups have produced profound changes in
the mechanical properties of the polymers. Unlike the silicones, however, the phosphazenes are soluble in a number of
organic solvents, providing a more simplified approach to
the making of blends, composites and coatings.

Although the SILASTIC[R] silicone rubber sheeting lacks
the solubility of the organophosphazene polymers, it was
selected as the control because it has been used extensively

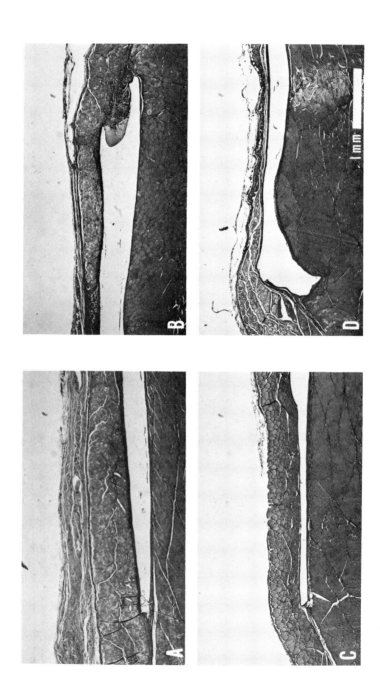

Fig. 2. Low magnification photomicrographs of implants between rat gluteal muscles showing typical thin fibrous capsules. Implants: SILASTIC®, A (7 days), C (28 days) and poly[bis(anilino) – phosphazene] B (7 days) and D (28 days).

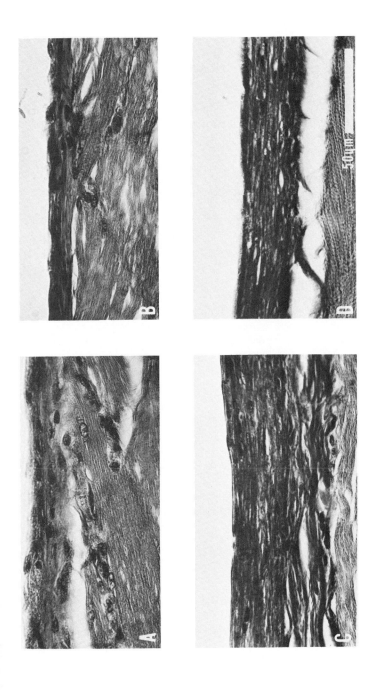

Fig. 3. High magnification photomicrograph of interfaces of rat gluteal muscle tissues and implants (clear areas) showing encapsulation of implant by a thin layer of fibrous tissue with minimal inflammatory and foreign body response and no damage to underlying gluteus medius muscle. Implants: SILASTIC®, A (7 days), C (28 days) and poly[bis(anilino)phosphazene], B (7 days) and D (28 days).

in reconstructive surgery and found acceptable for augmentation, space filling, and subdermal tissue replacement[20]. It is mechanically compatible with many of the soft tissues of the body and can be implanted without fear of adverse reaction to or from the tissues. Because of the many similarities in its chemical properties to those of the phosphazenes, it seemed to be a desirable control material.

The control material is but one of many variables for which standards are needed but have not been established. Consequently, particular attention was given to implant shape, animal species, implantation site, length of time for implantation and method of evaluation. The rationale used in their selection is briefly mentioned because of their importance in in vivo testing. We selected the rat because of its availability, because of the preliminary nature of the study and because approximately 40 animals were needed. In addition, the rat muscle was large enough to accommodate the implant specimens, whereas other sites, such as the abdominal cavity, mucous membrane and subcutaneous tissue were less appealing. Sterilization of the specimens was achieved with ^{60}Co gamma irradiation to prevent the problems associated with gas[21] and steam sterilization. Although we have left implants in animals for 1,3,7,14 and 28 days, our results have shown, as was observed by Turner, Autian and Lawrence[22] that the confounding effects of surgery are minimized after at least 7 days and of an acute toxic response after 28 days implantation. Hence, only two implant times were used. Sewell's quantitative method[18] for sutures seemed to be most suitable for this study and was used.

The facts that the specimens could be removed en bloc for sectioning and preparation, that there were no infections, and that the mild inflammatory response caused by SILASTIC^R was in excellent agreement with previous studies, suggest that the selected approaches were adequate for this study. Future studies may include some of these choices. Alternative procedures and problems associated with each are numerous and can be found in the literature[23, 24, 25].

Figs. 2 and 3 show the thin fibrous capsule and the cellular responses to the implants. The responses, whether to the poly(organophosphazenes) or to the silicone rubber are very similar and grossly are indistinguishable. The differences in the mechanical properties of the phosphazene polymers do not seem to have a significant role in inducing an inflammatory response. Also morphological properties, though responsible for some of the mechanical properties, do not appear to significantly influence the tissue response. Because there were no observable voids in the tissue, it was concluded that the brittle polymers remained intact in vivo.

Properties of the poly(organophosphazenes), except their biocompatibility, are more like the segmented poly(urethane)[26].
Where unique problems exist with urethane polymers, the
organophosphazenes may be suitable replacements.

In conclusion, the biocompatibility testing of eight
poly(organophosphazenes) has shown them to be as mild in rat
muscle tissue as is SILASTIC[R] silicone rubber. This and the
fact that the phosphazenes are soluble in certain organic
solvents from which they can be coated onto devices, cast into films or blended, warrant additional studies on their use
as materials for biomedical applications.

Acknowledgment. The authors thank Mr. John Roll for the rat
model drawing and Mr. Alfonzo Spencer for photographic repro-
ductions.

References

1.   Shaw, R.A., Fitzsimmons, B.W. and Smith, B.C., Chem.
     Rev., 62, 247(1962).
2.   Wilson, A., U.S. Army Natick Laboratories, Technical Re-
     port 75-38-CE, October 1974.
3.   Reynard, K.A., Woo, J.T.K., and Rose, S.H., Horizons,
     Inc., Cleveland, Ohio, AMMRC CR 71-17, Oct 1971
     (AD 734347).
4.   Reynard, K.A., Sicka, R.W., and Rose, S.H., Horizons, Inc.
     Cleveland, Ohio, AMMRC CTR 72-8, June 1972 (AD 745900).
5.   Wilson, A., Army Natick Laboratories, Technical Report,
     TR 75-38-CE, Oct 1974.
6.   Legally, P. and Klemens, W., Naval Ship Research and
     Development Center, MAT-74-42, Oct 1974.
7.   Rose, S.H. and Reynard, K.A., U.S. Patent 3,702,833(1972).
8.   Reynard, K.A. and Rose, S.H., Horizons, Inc., Cleveland,
     Ohio, NAS8-25184, June 1972 (N73-15593).
9.   Allcock, H.R., Chem. Rev., 72, 315(1975).
10. Allcock, H.R., "Phosphorus-Nitrogen Compounds", Academic
     Press, New York (1972).
11. Allcock, H.R. and Kugel, R.L., J. Am. Chem. Soc. 87,
     4216(1965).
12. Allcock, H.R., Kugel, R.L. and Valan, K.J., Inorg. Chem.
     5, 1709(1966).
13. Allcock, H.R., Cook, W.J. and Mack, D.P., Inorg. Chem.,
     11, 2584(1972).
14. Singler, R.E., Schneider, N.S. and Hagnauer, G.L.,
     Polymer Eng. Sci., 15, 321(1975).
15. White, J.E., Singler, R.E. and Leone, S.A., J. Polym.
     Sci., Polym. Chem. 13, 2531(1975).
16. Hagnauer, G.L. and Schneider, N.S., J. Polym. Sci. A-2,
     10, 699(1972).

17. Allcock, Harry R., Smeltz, Karen M., and Mack, Daniel P., U.S. Patent 3,893,980(1975) to Pennsylvania Research Corporation.
18. Sewell, W.R., Wiland, J. and Craver, B.N., Surg. Gynec. & Obstet., 100, 483(1955).
19. Snedcor, G.W. and Cochran, W.C., Statistical Methods, 6th edition, Iowa State University Press, Ames, Iowa, 1A, 1967.
20. Braley, S., Ann. N.Y. Acad. Sci., 146, 148(1968).
21. Kulkarni, R.K., Bartak, W., Ousterhout, D.K. and Leonard, F., J. Biomed. Mater. Res. 2, 165(1968).
22. Turner, J., Lawrence, W.H., and Autian, J., J. Biomed. Mater. Res. 7, 39(1973).
23. Matlaga, Barbara F., Yasenchak, Lewis P., and Salthouse, Thomas N., J. Biomed. Mater. Res. 10, 391(1976).
24. Oppeheimer, B.S., Oppenheimer, E. and Stout, A.P., Proc. Soc. Exp. Biol. Med. 79, 366(1952).
25. The Pharmacopeia of the United States of America, XVIII edition, Mack Publishing Company, Easton, Pennsylvania, 1970, pp. 926-929.
26. Hicks, E.M., Ultee, A.J., and Drougas, J., Science 147, 373(1965).

POLYMERIC SULFUR NITRIDE, $(SN)_x$,
AND ITS HALOGEN DERIVATIVES

M. Akhtar, C.K. Chiang,[*] M.J. Cohen,[*] A.J. Heeger,[*]
J. Kleppinger, A.G. MacDiarmid, J. Milliken,
M.J. Moran and D.L. Peebles[*]
Department of Chemistry and
[*]Department of Physics,
University of Pennsylvania

Monoclinic crystals of metallic $(SN)_x$ may be synthesized by the solid state polymerization of S_2N_2 at room temperature. The monoclinic form may be converted to an orthorhombic form by mechanical shearing. When monoclinic $(SN)_x$ is heated at $145°$ it depolymerizes to a gaseous, very reactive isomer of S_4N_4 which re-polymerizes on room temperature glass or plastic substrates to give golden, metallic $(SN)_x$ films. Crystals of $(SN)_x$ have a large conductivity at room temperature which shows the temperature dependence characteristic of metals, increasing rapidly with decreasing temperature. When $(SN)_x$ is treated with bromine vapor at room temperature it is converted to shiny black crystals of $(SNBr_{0.55})_x$ which lose bromine on pumping at room temperature to give compositions in the range $(SNBr_y)_x$ ($y \sim 0.3$-0.4). Simultaneous heating and pumping yields copper-colored crystals of $(SNBr_{0.25})_x$. All the brominated species are apparently metals having a higher conductivity than $(SN)_x$. Conductivity and optical reflectance measurements strongly suggest that the brominated polymers are more anisotropic in their metallic properties than $(SN)_x$.

Metals, by definition, are those elements which possess certain characteristic chemical properties such as ease in forming positive ions by chemical processes and certain characteristic solid state physical properties such as high electrical conductivity (which increases as the temperature is lowered), high reflectivity of light, good thermal conductivity, ductility, and malleability. Until very recently, it was believed that these collective properties were unique

to metallic elements, but it now appears that many of these
properties may be found in a simple inorganic covalent polymer
that contains no metal atom. Studies of the compound, polymeric sulfur nitride (polythiazyl), $(SN)_x$, reveal that it possesses a sufficient number of the above properties to be termed a
metal. It is believed that this polymer may be the first
example of a new class of material - polymeric molecular metals. Indeed, very recent studies involving the synthesis of
a series of metallic derivatives of $(SN)_x$, the polythiazyl
bromides, $(SNBr_y)_x$, have given support to this concept.

Polymeric sulfur nitride is prepared conveniently in the
form of lustrous, golden monoclinic crystals by the solid
state polymerization of molecular crystals of disulfur dini-
tride, S_2N_2, at room temperature (1). It may also be prepared
in the form of golden epitaxial films, in which the $(SN)_x$
chains are aligned parallel to each other (2). The polymer
has electrical and optical properties characteristic of an
anisotropic metal, the metallic properties being more pronounced along the direction of the $(SN)_x$ polymer chain than
in directions perpendicular to the chain (3). It was first
synthesized in 1910 by Burt (4), who passed the vapor of S_4N_4
(which may be prepared in a two-step reaction from sulfur,
chlorine and ammonia) over silver gauze or quartz wool at
100-300°C. In 1956 M. Goehring (5) prepared S_2N_2 from S_4N_4
vapor and silver wool at 300°C and found that the S_2N_2 polymerized spontaneously in the solid state to $(SN)_x$ at room
temperature. In 1973, Labes (6) reported measurements of
electrical properties of crystalline bundles of $(SN)_x$ fibers.
The dc electrical conductivity exhibited room temperature
values ranging from (0.01 to 1.7) x 10^3 (ohm cm)$^{-1}$ and a
negative temperature coefficient (increase of conductivity
with a decrease in temperature), suggestive of metallic behavior. For comparison, the room temperature conductivity of
a metal such as mercury is 10.4 x 10^3 (ohm cm)$^{-1}$. Studies by
two groups of investigators, A.F. Garito, A.J. Heeger and
A.G. MacDiarmid (University of Pennsylvania) (7) and R.L.
Greene and G.B. Street (I.B.M. Corporation) (8) in 1974-75
confirmed that the metallic behavior of $(SN)_x$ is an intrinsic
property of the polymer.

Very recently an orthorhombic crystalline form of $(SN)_x$,
which has not yet been studied in detail, has been prepared
by Baughman et al. (9) from crystalline monoclinic $(SN)_x$ by
mechanical shear. The information given in this article
refers only to the monoclinic form of $(SN)_x$.

During the solid state polymerization of S_2N_2 crystals
at room temperature, they at first become dark blue-black and
paramagnetic and then change to lustrous golden diamagnetic

$(SN)_x$ crystals. The polymerization process is depicted diagramatically in Fig. 1 where one of the bonds in the square
planar S_2N_2 molecules spontaneously breaks upon thermal activation to give the paramagnetic species (1). The bond angles
widen and the units join together to give the $(SN)_x$ polymer
chain, a section of which is shown at the bottom of Fig. 1.

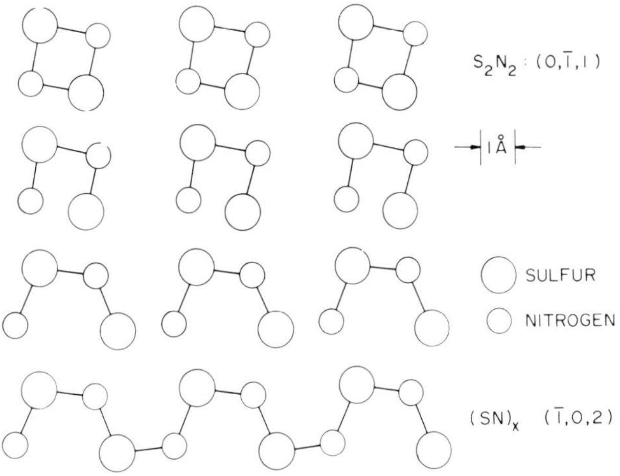

$S_2N_2 : (0,\bar{1},1)$

$\dashv$ 1 Å $\vdash$

◯ SULFUR

◯ NITROGEN

$(SN)_x$ $(\bar{1},0,2)$

*Fig. 1. The polymerization of S_2N_2 to $(SN)_x$. The top
view is a projection of the S_2N_2 structure onto the $(0\bar{1}1)$
plane with the a axis horizontal. The bottom view is a projection of the $(SN)_x$ structure onto the $(\bar{1}02)$ plane with the
b axis horizontal. The middle views schematically show the
polymerization process.*

The fibrous nature of a portion of an imperfect polymer
crystal may be seen from the scanning electron micrograph
shown in Fig. 2. A well-formed $(SN)_x$ crystal is shown in
Fig. 3.

Single crystal x-ray studies of $(SN)_x$ (1) show that $(SN)_x$
forms as planar, parallel $(SN)_x$ chains in a monoclinic crystalline lattice. The S-N bond lengths in the polymer are intermediate between those expected for a single and double S-N
bond.

Fig. 2. Scanning electron microscope photograph of an imperfect crystal illustrates the fibrous nature of polymeric sulfur nitride.

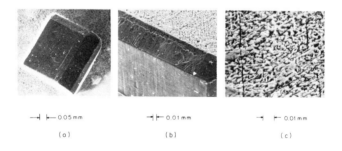

Fig. 3. Scanning electron micrographs of a crystal of $(SN)_x$; (a) a face; (b) the face, the upper right hand edge and the end (not visible in (a)); (c) the end shown in (b) showing clearly the $(SN)_x$ fiber ends.

From simple valence bond concepts, the bonding between sulfur and nitrogen atoms in an $(SN)_x$ chain may therefore be considered as derived, to a first approximation, from the two extreme resonance forms

in which all atoms exhibit normal oxidation states and valences. These give the resonance hybrid species

in which all bonds are intermediate between double and single bonds.

Monomeric NS, the sulfur analog of NO, has a transitory existence in the gaseous state. Like NO, it has one unpaired electron in the lowest energy antibonding pi molecular orbital. In a single $(SN)_x$ chain, overlap of these half-filled π^* orbitals provides the principal basis for the electron delocalization into the metallic conduction band. Calculations based on the x-ray data (10) suggest that there are weak but important S-S, N-N and N-S interactions between different SN chains. A diagrammatic representation of one of these interchain interactions involving sulfur atoms in $(SN)_x$ chains lying parallel to the ($\bar{1}02$) crystallographic plane of $(SN)_x$ is illustrated in Fig. 4 (dashed lines). It is interactions of this type which are probably responsible for $(SN)_x$ maintaining its metallic properties down to very low temperatures.

When $(SN)_x$ is depolymerized at ca 145°C, it is converted to a gaseous, very reactive isomer of S_4N_4 (2,11). When this vapor impinges on room temperature, glass, plastic or metal surfaces, spontaneous repolymerization occurs to give golden $(SN)_x$ films having the same crystal structure as the $(SN)_x$ crystals from which they were derived (2). If the room temperature surface consists of, for example, a polymer such as Mylar or Teflon which has previously been aligned by appropriate stretching, then epitaxial films of $(SN)_x$ are deposited in which the $(SN)_x$ fibers all lie parallel to each other and to the direction of alignment of the polymer substrate (2).

The room temperature conductivity of good quality $(SN)_x$ crystals can be as high as approximately 3.7×10^3 (ohm cm)$^{-1}$. The conductivity has been observed to increase over 200 times on lowering the temperature to 4.2 K (12) and near 0.3 K the crystals become superconducting (13). The polymer is therefore

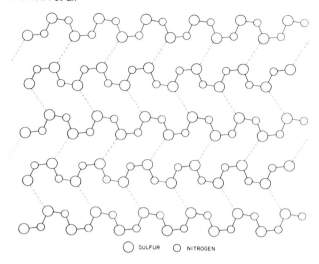

○ SULFUR ○ NITROGEN

Fig. 4. Diagrammatic representation of (SN)$_x$ chains parallel to the ($\bar{1}$02) plane of an (SN)$_x$ crystal. Sulfur-sulfur interactions are depicted by dashed lines.

the first example of a material consisting of strictly non-metallic elements to exhibit superconductivity at 1 atm pressure.

The reaction of (SN)$_x$ with bromine is particularly inter-esting and results in a new series of polymeric metals – the polythiazyl bromides, (SNBr$_y$)$_x$ (where y ∿ 0.25–0.55). Bernard et al. (14) have reported recently that (SN)$_x$ reacts with bromine vapor at room temperature to give a gray–blue solid the composition of which was not determined. Independent studies carried out by the University of Pennsylvania (15) and IBM (16) groups involved the characterization of crystals and films of (SNBr$_{0.4}$)$_x$, the first derivative of (SN)$_x$.

Samples of polythiazyl bromides, (SNBr$_y$)$_x$, were prepared by reacting analytically pure (SN)$_x$ with bromine vapor. For example, crystals of (SN)$_x$ (ca 1–10 mm^3) in 60 torr of bro-mine vapor react immediately at room temperature to give shiny black crystals having a lustrous blue–purple tinge. After pumping for approximately 30 minutes the material was analyzed. Compositions varied slightly, but significantly, according to minor variations in experimental procedure. Values obtained for (SNBr$_y$)$_x$ were in the range 0.38 ⩽ y ⩽ 0.42. Two typical analyses are given in Table 1. For convenience, this material will be referred to as (SNBr$_{0.4}$)$_x$ in this paper. No unreacted (SN)$_x$ could be detected visually or in the x-ray powder patterns of the material. Crystals of $(SNBr_{0.4})_x$ appear to react with air to a negligible extent during one hour at room temperature but are tarnished after 10 hours during which time some ammonium bromide is formed.

TABLE 1

Representative Elemental Analyses
of Selected Polythiazyl Bromides[a]

	%S	%N	%Br	Total
Calc. for $(SNBr_{0.40})_x$ [b]	41.09	17.95	40.96	100.00
Found	40.92	17.93	41.01	99.86
Calc. for $(SNBr_{0.38})_x$ [b]	41.95	18.32	39.73	100.00
Found	42.10	18.15	39.85	100.10
Calc. for $(SNBr_{0.25})_x$ [c]	48.54	21.21	30.25	100.00
Found	48.34	21.06	30.25	99.65

[a] Analyses performed by Galbraith Laboratories, Inc., Knoxville, Tennessee 37921.
[b] Shiny, black crystalline form of $(SNBr_y)_x$.
[c] Copper-colored crystalline form of $(SNBr_y)_x$.

Other compositions besides $(SNBr_{0.4})_x$ can also be synthesized. Crystals which were weighed under 60 torr of bromine vapor had a bromine content as high as $(SNBr_{0.55})_x$. When crystals of $(SNBr_{0.4})_x$ were pumped at room temperature for 45 hours or more, there was little change in appearance, but the bromine content decreased to as low as $(SNBr_{0.33})_x$. When crystals of $(SNBr_{0.4})_x$ were heated with pumping at approximately 86°C for about 30 hours they were converted to copper-colored crystals with composition $(SNBr_{0.25})_x$. (see Table 1)

The density of the $(SNBr_{0.4})_x$ (2.67 ± 0.04 g/cm^3) is significantly greater than that of $(SN)_x$ (2.30 g/cm^3). The volume increase calculated from the formula and density data is 46%, in reasonable agreement with the observed changes in crystal dimensions upon bromination. There is no significant change in the length of the crystal in the b crystallographic direction (the direction parallel to the $(SN)_x$ polymer chains). The entire volume change results from a swelling of the crystal perpendicular to the chain axis. Although the crystals appear well-formed, x-ray studies show they have very considerable defect structure. The fibrous nature characteristic of $(SN)_x$ crystals when mechanically pulled apart is preserved in $(SNBr_{0.4})_x$.

To monitor the kinetics of the chemical reaction and provide definitive data on the change in electrical resistance on bromination, the resistance was measured along the chain axis during the bromination of a single crystal of $(SN)_x$. The resulting normalized resistance vs. time during the reaction is shown in Fig. 5. The data were obtained in bromine vapor at a pressure of 20–25 torr. Chemical analysis of crystals brominated in the same experiment showed the reaction had proceeded to give the composition $(SNBr_{0.40})_x$.

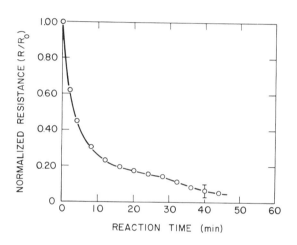

Fig. 5. Normalized resistance of $(SNBr_y)_x$ during bromination. The sample conductivity was 1.6×10^3 $(\Omega-cm)^{-1}$ before reaction increasing to 2.2×10^4 $(\Omega-cm)^{-1}$ after reaction.

The initial reaction proceeds rapidly and results in an obvious decrease in the sample resistance. The resistance eventually falls to a value approximately 0.05 of the initial value as shown in Fig. 5. Taking into account the ∿46% increase in cross-sectional area, a final room temperature conductivity approximately 13 times the initial $(SN)_x$ value is obtained. Independent measurements on five samples of $(SNBr_{0.4})_x$ mounted for measurement after reaction yielded values from 1.4×10^4 to 9.1×10^4 $(\Omega-cm)^{-1}$ with an average room temperature conductivity, in the direction parallel to the $(SN)_x$ chain, $\sigma_{\parallel}$ (300 K), equal to 3.8×10^4 $(\Omega-cm)^{-1}$. This may be compared with the value of 4.5×10^4 $(\Omega-cm)^{-1}$, the conductivity of lead at room temperature. The room temperature conductivity of $(SNBr_{0.4})_x$ perpendicular to $(SN)_x$ chain direction, $\sigma_{\perp}$, was 8 $(\Omega-cm)^{-1}$. However, because of the fibrous nature of the crystals, the value for σ_L is probably not intrinsic but is determined by inter-fibre contact.

The temperature dependence of the resistivity of $(SNBr_{0.4})_x$ along the chain axis is shown in Fig. 6. The data were obtained using four-probe dc techniques with Electrodag contacts. The normalized results for two samples shown in Fig. 6 indicate metallic behavior over the entire temperature range T < 300 K. Sample 1 (closed circles) had a room temperature conductivity of 9.1×10^4 $(\Omega\text{-cm})^{-1}$ which increased (approximately 90-fold) in a manner typical of metals on lowering the temperature to 4.2 K. Sample 2 (open circles) had a room temperature conductivity of 1.4×10^4 $(\Omega\text{-cm})^{-1}$ which increased approximately 12 times on lowering the temperature to 4.2 K.

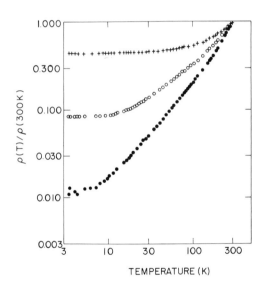

Fig. 6. Temperature dependence of the resistivity of polythiazyl bromides; ●●● $(SNBr_{0.4})_x$, ooo $(SNBr_{0.4})_x$, and +++ copper-colored $(SNBr_{0.25})_x$.

The temperature dependence of the resistivity of a crystal of copper-colored $(SNBr_{0.25})_x$ is also shown on Fig. 6 for comparison. The results again indicate metallic behavior. Measurements on six samples of $(SNBr_{0.25})_x$ yielded an average room temperature conductivity of 2×10^4 $(\Omega\text{-cm})^{-1}$ with values ranging from 0.9×10^4 to 5.9×10^4 $(\Omega\text{-cm})^{-1}$.

Much important information concerning the electronic properties of a metal can be obtained from its reflectance spectrum. Reflectance measurements were therefore carried out from the face of a single crystal of $(SNBr_{0.4})_x$ using polarized light (see Fig. 7). The curve from 1.0 to 2.5 eV for light polarized parallel to the SN chains is characteristic of that of a metal. The featureless spectrum of light polarized perpendicular to the SN chains is not characteristic of a metal and suggests that $(SNBr_{0.4})_x$ is more anisotropic in its metallic properties than $(SN)_x$.

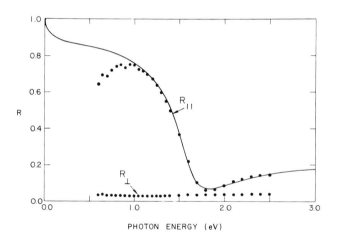

PHOTON ENERGY (eV)

Fig. 7. Polarized reflectance of $(SNBr_{0.4})_x$ in the spectral range 0.6 eV to 2.5 eV. The solid curve represents a least-square fit arbitrarily restricting the input data to the range above 1.1 eV. The data are plotted using the absolute normalization as determined from the Drude fits.

Visual observation of the crystals after reaction suggests that bromination decreases the crystalline quality. X-ray studies confirm these latter observations; the relatively well-defined x-ray patterns obtained from $(SN)_x$ crystals become broad and diffuse indicative of a significant decrease in crystal quality. An analysis of the conductivity and optical data provides evidence that the fundamental $(SN)_x$ chains remain intact as bromine enters the structure with little or no charge transfer and suggest an analogy with bromine intercalation compounds of graphite (17). That at least a portion of the bromine is only weakly bound is evident from the observation that significant amounts of it may be removed by pumping $(SNBr_{0.4})_x$ at room and/or at slightly elevated temperatures.

The isolation of this series of metallic derivatives of $(SN)_x$ strongly suggests that it may be possible to synthesize relatively easily a wide variety of other derivatives which are also metallic.

ACKNOWLEDGEMENT

This work was supported by the National Science Foundation by MRL program grant DMR 76-00678 and by grant GP-41766X and by the Office of Naval Research contract No. N00014-75-C-0962.

REFERENCES

1.   Mikulski, C.M., Russo, P.J., Saran, M.S., MacDiarmid, A.G., Garito, A.F., and Heeger, A.J., J. Amer. Chem. Soc. 97, 6358 (1975); Cohen, M.J., Garito, A.F., Heeger, A.J., MacDiarmid, A.G., Mikulski, C.M., Saran, M.S., and Kleppinger, J., J. Amer. Chem. Soc. 98, 3844 (1976).
2.   Bright, A.A., Cohen, M.J., Garito, A.F., Heeger, A.J., Mikulski, C.M., and MacDiarmid, A.G., Appl. Phys. Lett. 26, 612 (1975).
3.   Geserich, H.P., and Pintschovius, L., in Festkörperprobleme (Advances in Solid State Physics), (J. Treusch, Ed.), Vol. XVI, p. 65. Vieweg, Braunschweig, 1976; Street, G.B., and Greene, R.L., IBM Res. Develop. 21, 99 (1977).
4.   Burt, F.P., J. Chem. Soc. 1171 (1910).
5.   Goehring, M., and Voight, D., Z. Anorg. Allg. Chem. 285, 181 (1956).
6.   Walatka, V.V., Jr., Labes, M.M., and Perlstein, J.H., Phys. Rev. Lett. 31, 1139 (1973).
7.   Bright, A.A., Cohen, M.J., Garito, A.F., Heeger, A.J., Mikulski, C.M., Russo, P.J., and MacDiarmid, A.G., Phys. Rev. Lett. 34, 206 (1975).
8.   Greene, R.L., Grant, P.M., and Street, G.B., Phys. Rev. Lett. 34, 89 (1975).
9.   Baughman, R.H., Apgar, P.A., Chance, R.R., MacDiarmid, A.G., and Garito, A.F., J. Chem. Phys. 66, 401 (1977).
10. See Ref. 1 and references therein.

11. Smith, R.D., Wyatt, J.R., DeCorpo, J.J., Saalfeld, F.E., Moran, M.J., and MacDiarmid, A.G., J. Amer. Chem. Soc. 99, 1726 (1977).

12. Chiang, C.K., Cohen, M.J., Garito, A.F., Heeger, A.J., Mikulski, C.M., and MacDiarmid, A.G., Solid State Commun. 18, 1451 (1976).

13. Greene, R.L., Street, G.B., Suter, L.J., Phys. Rev. Lett. 34, 577 (1975).

14. Bernard, C., Herold, A., Lelaurain, M., and Robert, G., C. R. Acad. Sc. Paris C, 283, 625 (1976).

15. Akhtar, M., Kleppinger, J., MacDiarmid, A.G., Milliken, J., Moran, M.J., Chiang, C.K., Cohen, M.J., Heeger, A.J., and Peebles, D.L., Chem. Commun., in press.

16. Street, C.B., Gill, W.D., Geiss, R.H., Greene, R.L., and Mayerle, J.J., Chem. Commun., in press.

17. Platts, D.A., Chung, D.D.L., and Dresselhaus, M.S., Phys. Rev. B15, 1087 (1977).

COORDINATION POLYMERS AND THEIR USES

John C. Bailar, Jr.
University of Illinois at Urbana-Champaign

ABSTRACT

A coordination polymer is one which a metal ion
is coordinated with an organic substance so as to form
a polymer in which there is a metallic atom or ion for
each monomer unit. There has been little work on
them, but several uses have been developed.

Coordination polymers have been known and used for many
years, though they have not been recognized as such until
fairly recent times. The tanning of leather, which is an
ancient art, depends upon the coordination of metal ions with
the polypeptides of hide. The complexes which are thus
formed are much more resistant to bacterial attack, weather
and wear than is the original hide. Sodium metaphosphate,
which is a long chain polymer, has long been used as a water
softener because it coordinates with calcium, magnesium, fer-
rous and other metal ions.
Palladium(II) chloride and nickel cyanide are respectively linear and sheet polymers and the heavy metal ferro-
and ferricyanides, in which the iron ions show a coordination
number of six, are three dimensional, infinite polymers.
It has long been known that the hydroxy and carboxyl
groups readily form "bridges" between metal ions, thus forming polymeric species (1). This property has been used in
the formation of a water repellent "basic chromic stearate,"
a "basic chromic acrylate" and an analogous hydrophobic,
oleophobic perfluorocarboxylate. The monomers of these sub-
stances are soluble in the lower alcohols, and are applied to
the surfaces to be protected in that form. When the material
is heated, the formation of hydroxo and carboxylate bridges
takes place, binding the chromium ions firmly together with
hydrocarbon or perfluorocarbon chains protruding from the

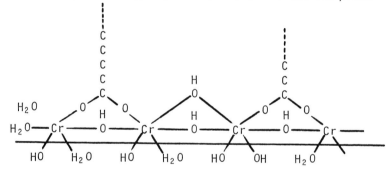

$$[Cr(H_2O)_6]^{3+} + OH^- \rightarrow$$

surface and protecting it.   In the case of the acrylate

chains, the protection is increased by polymerizing other
groups to the acrylate. (In actual practice, there are always some chloro groups in these protecting films. They are
not shown in the schematic diagram.)

Only in recent times have "Werner complexes," as we
commonly think of them, been designed to form polymers which
would find industrial use. An early example is furnished by
the polarizing films which were patented by Amon and Kane
(2). According to their procedure, a sheet of an organic
plastic is impregnated with a solution of a rubeanate (a
dithiooxamide) and then with a solution of a copper or nickel
salt. The rubeanate ion and the metal ion react in the plastic sheet to form long chain complexes. These chains become
oriented when the plastic is stretched in one direction.

During World War II, the U. S. Air Force sponsored a
large research program in attempts to find polymers that
would be stable up to 600°C. The thought that coordination

$$Me^{2+} + \quad \begin{matrix} NR{=\!=\!=}C{-\!-}SH \\ | \\ NR{=\!=\!=}C{-\!-}SH \end{matrix} \quad \longrightarrow \quad Me \begin{matrix} {-}NR{=\!=\!=}C{-}S \\ | \\ NR{=\!=\!=}C{-}S \end{matrix} Me \begin{matrix} {-}NR{=\!=\!=}C{-}S \\ | \\ NR{=\!=\!=}C{-}S \end{matrix} -$$

might give extremely stable polymers probably originated in the knowledge that simple organic compounds are greatly stabilized by coordination with metal ions. Thus, N-hydroxy-ethylethylenediamine is quickly destroyed by hot nitric acid, but the complex

$$\left[Co{\Huge(}\begin{matrix} NH_2 - CH_2 \\ NH - CH_2 \\ CH_2CH_2OH \end{matrix}{\Huge)}_3 \right]^{3+}$$

can be heated with concentrated nitric acid for many hours without decomposition (3). Similarly, the bis ketoimine

$$\begin{matrix} H_3C \\ \\ H_2C \\ \\ H_3C \end{matrix} \begin{matrix} C{=\!=}O \quad\quad O{=\!=}C \\ \\ C{=\!=}NCH_2CH_2N{=\!=}C \end{matrix} \begin{matrix} CH_3 \\ \\ CH_2 \\ \\ CH_3 \end{matrix}$$

decomposes at moderate temperatures, but its copper complex is decomposed only slowly at 350°C (4).
    Unfortunately, the Air Force search for high temperature coordination polymers did not yield such substances. It was found that, in most cases, only short chains were formed, and also that thermally stable "monomers" lost much of their stability when linked into polymeric units. This was prob-ably due to breakdown of the organic groups which connected the metal ions together. If coordinate bonds do lend sta-bility, it can extend only to the few atoms which are rea-sonably close to the metal. Unfortunately, if the connect-ing chains are short enough to be stable, they lack flexi-bility.
    Perhaps the closest approach to the standard set by the Air Force was achieved by Martin and Judd (5), who synthe-sized derivatives of several thiopicolinamids. For example, the zinc derivative of

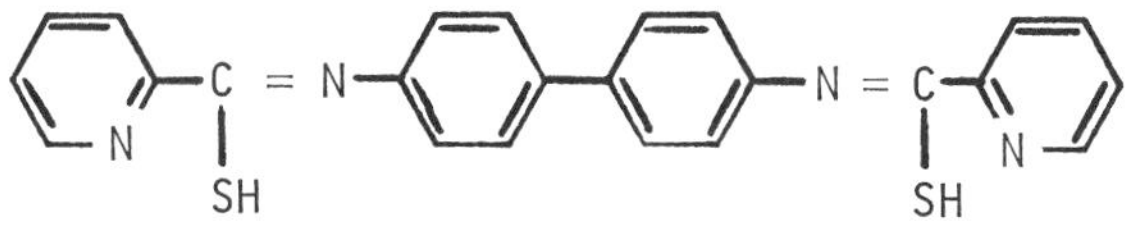

was found to be stable at 400°C in the absence of air.

There are three general ways in which coordination polymers can be prepared:

1. By the use of a ligand which can attach itself simultaneously to two metal atoms or ions. Many such ligands have been studied, most of them being bis-chelating agents such as bis-1,3-diketones, bis-8-hydroxyquinolines, and rubeanates. When ligands which are not chelaters are used, "double bridges" are usually formed:

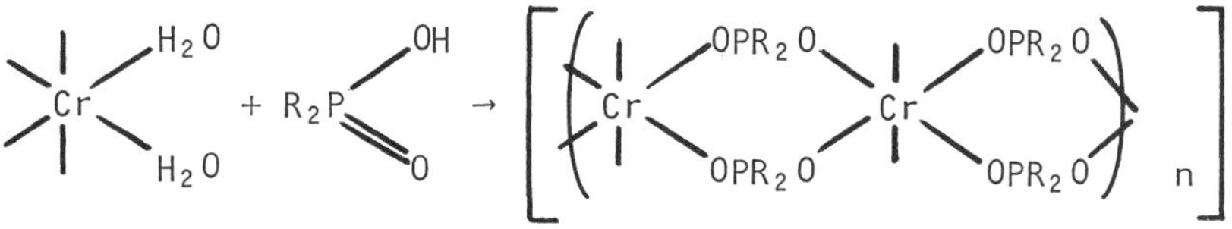

Block and his coworkers (6) have prepared a large number of polymers (and have published many papers) in which metal ions are linked together by substituted phosphinous anions

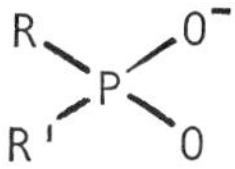

. In Block's examples, R and R' are usually the same

and are either phenyl or methyl. The score or more of metal ions which Block has used vary in oxidation state and coordination number (7), so a great many polymers are possible. The simplest are linear polymers which can be formulated as

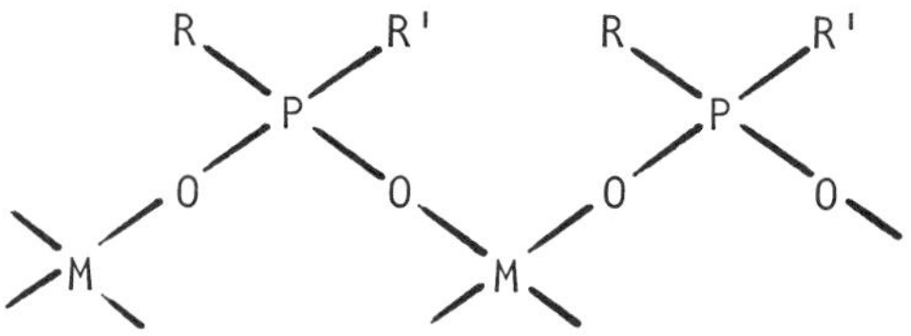

but many other types exist. Some of these polymers have high molecular weights (up to 150,000), some are thermally stable, some are soluble in organic solvents, some are fusible, some can be cast into films, and so on. Several have been suggested as viscosity stabilizers for oil and grease thickeners (8), some as anti-static agents (8) and some as corrosion resistant coatings for metals (9).

2. By the use of a coordination compound which has, on

its periphery, groups which can be used in polymer formation, such as

$$
\begin{array}{c}
\text{HO}-\text{C} \overset{R}{\underset{R}{\Longleftarrow}} \text{C}=\text{O} \quad \text{O}=\text{C} \overset{R}{\underset{R}{\Longrightarrow}} \text{C}-\text{OH} + \text{OCN}-\text{R}'-\text{NCO} \longrightarrow
\end{array}
$$

$$(10)$$

$$
\left(-\text{C} \overset{R}{\underset{R}{\Longleftarrow}} \text{C}=\text{O} \quad \text{O}=\text{C} \overset{R}{\underset{R}{\Longrightarrow}} \text{C}-\text{NH}\overset{\text{O}}{\overset{\|}{\text{C}}}-\text{R}'-\overset{\text{O}}{\overset{\|}{\text{C}}}-\text{HN}- \right)_n
$$

In place of diisocyanates, one may use $SOCl_2$, SO_2Cl_2, or other compounds that react readily with active hydrogen atoms.

  3.   By the use of a preformed organic or inorganic polymer to which metal atoms or ions can be chemically bound:

$$
\text{OHC}-\bigcirc-\text{SO}_2-\bigcirc-\text{CHO} + \bigcirc\overset{}{\underset{H_2N \quad NH_2}{}} \longrightarrow
$$

$$
\left(=\text{HC}-\underset{HO}{\bigcirc}-\text{SO}_2-\underset{OH}{\bigcirc}-\text{CH}=\text{N} \quad \text{N}=\text{CH}-\underset{OH}{\bigcirc}-\text{SO}_2-\underset{OH}{\bigcirc}-\text{CH}= \right)_n \longrightarrow
$$

$$
\left(=\text{HC}-\underset{HO}{\bigcirc}-\text{SO}_2-\underset{O}{\bigcirc}-\text{CH}=\text{N} \quad \text{N}-\text{CH}-\underset{O}{\bigcirc}-\text{SO}_2-\underset{OH}{\bigcirc}-\text{CH}= \right)_n
$$

where M is a dipositive four covalent metal ion (11).

  Polymers to which metal ions and complexes are bound are finding use in "heterogenizing" homogeneous catalysts. If a homogeneous catalyst is attached to a polymer which is insoluble in the reaction medium, it may still retain the selectivity of the homogeneous catalyst, but like other heterogeneous catalysts, it can be recovered readily by filtration. Biologists have known for many years that enzymes, which catalyze reactions in living systems, are bound to cell walls and other membranes (12). An enzyme thus held in place can undergo reaction many times before it is carried out of the reaction zone. The N.S.F. is subsidizing research on the binding of enzymes to insoluble materials such as cellulose and glass. Perhaps these catalysts can bring about the conversion of cellulose and hydrocarbons to food stuffs. Other applications are also under study (13). It is perhaps surprising that it took so long for chemists to realize that synthetic homogeneous catalysts could be bound to polymers and thus rendered heterogeneous. One of the earliest applications of this concept was that of Haag and Whitehurst at Mobil Oil (14), who attached Group VIII metal complexes to polystyrene which was cross-linked by divinyl benzene.

The attachment of catalysts to polymers has now been widely studied, and many variations have been developed. Most commonly, phosphines have been used to coordinate with the metal, but other donor groups are effective, too (15). Catalysts bound to polymers do not lose selectivity, and in some cases are more active than the parent homogeneous catalysts. Polymer-rhodium carbonyl complexes are effective and highly selective in the hydroformylation of olefins (16). Bruner and Bailar (17) used polystyrene-divinyl-benzene-phosphine polymers with coordinated platinum(II) chloride or palladium chloride in the selective hydrogenation of soybean methyl ester.

Allyls, vinyls, urethanes and many other polymers can be as effective as polystyrene, and have been widely used for hydrogenations (15,16), carbonylations (18) and oligomerizations (15).

Some remarkable variations have styrene tricarbonyl chromium (19) and titanocene dichloride (20) appended to the polystyrene chain. The latter was found to be six times as as active as the monomer.

Even silica has been used as the polymer. The silanol groups on the surface of the silica are made to combine with reactive materials such as $SOCl_2$, $RSiCl_3$ or active alcohols. These, in turn, hold coordination groups which attach themselves to metal ion catalysts. An excellent example of this is shown in the equation:

$$(SiO_2)_n OH + X_3 Si(CH_2)_m PR_2 \rightarrow (SiO_2)_n OSiX_2(CH_2)_m PR_2$$

($m = 2, 6, 8, 10, 14$) ($X = $ -Cl, -OCOMe, -OMe, -OEt) ($R = $ cyclohexyl, phenyl, propyl). Attachment of the phosphine to rhodium carbonyl chloride, rhodium 1,5-cyclooctadiene chloride dimer and cobalt(III) acetylacetonate gives effective hydroformylation catalysts (21).

The stereochemistry of polymer attached catalysts is not known. Models of the polystyrene-phosphine-metal catalysts indicate that two adjacent phosphorus atoms can readily attach themselves to the trans positions of a planar platinum (II) or palladium(II) complex (22). If the polymer is coiled, two distant phosphorus atoms can attach themselves to a single metal atom, or the metal can combine with phosphorus atoms in two different chains.

Mitchell and Whitehurst (23) have shown that quaternization takes place in some cases, and that the resultant phosphonium complexes are very poor catalysts.

REFERENCES

1.   Rollinson, C. L., in "The Chemistry of Coordination Compounds" (John C. Bailar, Jr., Ed.), Reinhold, New York, 1956, Chapter 13, summarizes the demonstrations of this theory up to 1956.
2.   Amon, W. F., Jr., and Kane, M. W., U. S. Patent 2,505,085 (April 25, 1950).

3. Keller, R. N., and Edwards, L. J., J. Am. Chem. Soc. 74, 215 (1952).
4. Charles, R. G., J. Phys. Chem., 65, 568 (1961).
5. Martin, K. V., and Judd, M. L., Report on Contract No. AF 33(616) 3200 (Nov. 1956) to U. S. Air Force, Wright Air Development Center, Final Report on Contract No. AF 33(616) 3772 (Dec. 1957).
6. Block, B. P. and his coworkers have published many papers and have patented a large number of these sub-stances. For reviews of their work, which is still underway, see U. S. Clearinghouse Fed. Sci. Tech. Inform., 1968, A. D. 678324, from U. S. Gov. Res. Develop. Rep., 1969, 9, 56 (1969); Inorg. Macromol. Rev., 1970, 1 (2) 115.
7. Coordination Chemistry (papers presented in honor of John C. Bailar, Jr.), Stanley Kirschner, Ed., Plenum Press, New York (1969).
8. Dahl, Gerd H., Sprout, Oliver S., Peschko, Norman D., Block, B. Peter, U. S. Clearinghouse Fed. Sci. Tech. Inform., 1968, A. D. 671,882, from U. S. Gov. Res. Develop. Rep., 1968, 68 (17) 103.
9. Block, B. P., and Dahl, Gerd, U. S. Patent 3,415,781 (Dec. 10, 1968).
10. Klein, Richard M., and Bailar, John C., Jr., Inorg. Chem. 2, 1190 (1963).
11. Marvel, C. S., and Tarkoy, N., J. Am. Chem. Soc. 79, 6000 (1957). See alos Goodwin, H. A., and Bailar, John C., Jr., J. Am. Chem. Soc. 83, 2467 (1961).
12. Bailar, John C., Jr., Catalysis Rev.-Sci. Eng. 10, 17 (1974); Manassen, J., in "Catalysis, Progress in Re-search," (F. Baslolo and R. L. Burwell, Jr., Eds.), p. 177, Plenum Press, New York, 1973.
13. Mosbach, K., Sci. Amer. 224, 26 (March, 1971).
14. Haag, W., and Whitehurst, D. D., Belgian Patent 721,686 (1968), German Offen. 1,800,379 (April 30, 1969).
15. Haag, W. and Whitehurst, D. D., Catalysis, Vol. 1, Proceedings of the 5th International Congress on Catalysis, Palm Beach, Fla., August, 1972 (J. W. Hightower, ed.), North Holland, New York, 1973, p. 29.
16. Jurewicz, A. T., Rollmann, L. D., and Whitehurst, D. D., Abstracts of papers, American Chemical Society Meeting, Chicago, Aug. 26-31 (1973), p. INDE 032.
17. Bruner, H. S., and Bailar, John C., Jr., J. Am. Oil Chem. Soc. 49, 533 (1972); Inorg. Chem. 12, 1465 (1973). For similar cases, see Ref. 13 and articles cited therein. There are many other examples.
18. Pittman, C. U., Jr., and Evans, G. O., Chem. Technol. 1, 416 (1971); Evans, G. O., Pittman, C. U., Jr.,

McMillan, R., Beach, R. T., and Jones, R., _J. Organo-metal. Chem._ 67, 295 (1974); Butter, S. A., U. S. Patent 3,700,702 (Oct. 24, 1972).

19. Pittman, C. U., Jr., Grube, P. L., Ayers, O. E., McManus, S. P., Rausch, M. D., and Moser, G. A., _J. Polym. Sci., Polym. Chem. Edit._, 10, 379 (1972).

20. Grubbs, R. H., Gibbons, C., Kroll, L. C., Bonds, W. D., Jr., and Brubaker, C. H., Jr., _J. Am. Chem. Soc._ 95, 2373 (1973).

21. Oswald, A. A., Murrell, L. L., and Boucher, L. J., Prepr., _Div. Pet. Chem. Am. Chem. Soc._, _1974_, 19 (1), 155; Boucher, L. J., Oswald, A. A., and Murrell, L. L., ibid. p. 162.

22. Bruner, H. S., Thesis, University of Illinois at Urbana (1971).

23. Mitchell, T. O., and Whitehurst, D. D., Third North American Conference of the Catalysis Society, San Francisco, Feb. 1974.

THALLIUM NMR AS A PROBE OF ALKALI ION INTERACTIONS AND OF COVALENCY IN ORGANOTHALLIUM COMPOUNDS

Jeffrey I. Zink, Chit Srivanavit, and James J. Dechter
Department of Chemistry
University of California, Los Angeles

The thallium(205) chemical shift is greater than 4800 ppm and is very sensitive to the environment of the nucleus. For thallous ions, the change of the chemical shift with solvent is used to measure the preferential solvation of the ion. In addition, it is used to measure the stability constants of the cations Li^+, Na^+, K^+, Rb^+, Cs^+, NH_4^+, Ag^+ with macrocyclic polyethers relative to that of the thallous ion. In organothallium compounds, the shift is a sensitive probe of covalency. Substituents on the phenyl ring in arylthallium compounds change the thallium chemical shift by over 55 ppm. The plot of the chemical shift vs. the substituent constant is roughly linear, but the shielding decreases with increasing electron donation in contrast to all other nuclei except ^{31}P and ^{199}Hg.

I. INTRODUCTION

Thallium is unique among the group IIIb metals in possessing stable +1 and +3 oxidation states. The most abundant isotope, ^{205}Tl, has a natural abundance of 70.48% and a spin of one half. The range of chemical shifts for ^{205}Tl which has been observed to date is about 2200 ppm for thallous salts in solution [1] and about 4800 ppm for thallic compounds [2]. Because of the large range of chemical shifts and their sensitivity to the environment of the thallium, ^{205}Tl nmr is very useful for probing such fundamental and diverse problems as ion pairing, preferential solvation of ions, biological cation selectivity, and covalency in organometallic compounds. In this paper, the applications of thallium nmr to ion solvation, ion binding by crown ethers, and electronic effects in alkyl and arylthallium compounds are discussed.

II. PREFERENTIAL SOLVATION

The chemical shift of the thallous ion in a mixed solvent system reflects the preferential solvation of the ion by
one of the solvents over the other as shown in figure 1 (3).

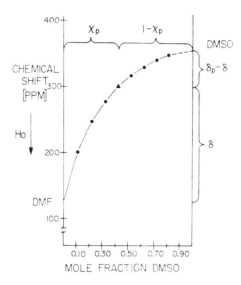

*Fig. 1. Mole fraction dependence of the chemical shift
for a DMF-DMSO mixture illustrating the calculation of $K^{1/n}$
at the point denoted by ▲.*

Two measures of preferential solvation are in current use.
The first, proposed by Popov (4) from alkali metal ion nmr,
is the isosolvation number which is defined as the mole fraction of a solvent B which must be mixed into a second solvent
A in order to produce an observed chemical shift midway between the chemical shifts in the two pure solvents. A second
measure, defined by Covington (5) for the equilibrium
$M(A)_n + nB = M(B)_n + n(A)$, is shown in equation 1. The quantity $1/n \log K$ is related to the free energy of solvation rela-

$$K^{1/n} = \left(\frac{\delta}{\delta p - \delta}\right)\left(\frac{1 - X_p}{X_p}\right) \qquad (1)$$

tive to that of a reference solvent. The other symbols are
defined in figure 1. We have shown that the two measures are

related (3).

Data of the type shown in figure 1 can be used to calculate 1/nlogK values for a variety of solvents. Representative values are given in table 1 in order of decreasing solvating

TABLE 1

Experimental Values of Relative Solvating Ability and the Tl^+ Chemical Shift for Selected Solvents

Solvent	(1/n) log K_{DMF}[a]	(1/n) log K_{py}[a]	Chem Shift[b]
Pyrrolidine	1.127	0.898	1757
n-Butylamine	1.048		1896
Hexamethylphosphoramide	0.987	0.334	443
Diethylamine	0.788		794
Dimethyl sulfoxide	0.560	0.279	369
Pyridine	0.161	0.000	664
Dimethylformamide	0.000	-0.114	124
Pyrrole	-0.175		-506[d]
Dimethoxyethane[c]	-0.215		-170[d]
Tributyl Phosphate[c]	-0.431	-0.108	-20[d]
Methanol[c]	-0.470		35[d]
Dioxane[c]	-0.496		-130[d]
Tetrahydrofuran[c]	-0.686	-0.458	-80[d]
Methyl acetate[c]	-0.686		-130[d]
Acetone[c]	-0.967	-0.717	-240[d]
Propylene carbonate[c]	-1.252		-470[d]

[a]Error ± 0.02. [b]Error ± 1 ppm except as indicated.
[c]Solvent will not dissolve $TlClO_4$. [d]Shift obtained from fitting procedure; estimated error ± 20 ppm.

ability of the solvents. These values correlate with both the chemical shifts of the thallous ions in the pure solvents and with the other measures of solvation such as the Drago E and C numbers (6) and the Gutman donor numbers (7).

III. POLYETHER CATION SELECTIVITY SEQUENCES AND THEIR SOLVENT DEPENDENCES

The macrocyclic polyethers have aroused much interest because of the high stability of the complexes they form with the alkali and alkaline earth metal ions (8-13). Because the complexing of alkali metal ions by crown ethers is highly selective, the crown ethers have been used as models for macrocyclic antibiotics believed to be responsible for alkali ion transport across membranes (14-16). The ion selectivity sequences are explained as a superposition of the effects shown in equation 2 where K is the stability constant, i.e., the

$$-RT \ln K = \Delta G_{bind} - \Delta G_{solv(M^+)} - \Delta G_{solv(L)} - \Delta G_{conf(L)} \quad (2)$$

equilibrium constant for complexing, and the other terms are respectively the free energies of metal-ligand bonds, metal ion solvation, ligand solvation, and ligand conformation changes (9).

The stability constants should be strongly affected by the solvent because of the terms involving solvation of the ion and ligand. Previous measurements of the binding constants as a function of solvent have been restricted primarily to water and methanol (15) because of the limitations in the methods of measurement. A knowledge of the direction and magnitude of the solvent effects is important for theoretical interpretations of ion selectivity and practical applications to synthesis and separations.

The selectivity sequence of crown ethers toward a series of cations is measured using ^{205}Tl nmr as the ratio of stability constant of Tl^+-crown to M^+-crown where M^+ represents a cation. For a 1:1 stoichiometry of both the Tl^+ and M^+ complexes, the overall competition for coordination is represented by equation 3 where I is the crown ether. Because the

$$Tl^+ - I + M^+ \rightleftharpoons Tl^+ + M^+ - I \quad (3)$$

activity coefficients are unknown, the quantities reported here are the concentration stability constant ratios and not the thermodynamic stability constant ratios. The relative stability constants are calculated from equation 4 where

$$K^M/K^{Tl} = \frac{P\{[I]_{tot} - (1 - P)[Tl^+]_{tot}\}}{(1 - P)\{[M^+]_{tot} - [I]_{tot} + (1 - P)[Tl^+]_{tot}\}} \quad (4)$$

$[I]_{tot}$ and $[M^+]_{tot}$ are the total concentrations of ionophore and of added alkali ion, respectively. The population of uncomplexed Tl^+, P, is calculated from the chemical shift information as shown in equation 5 where δ is the chemical shift at

$$P = (\delta - \delta_I)/(\delta_f - \delta_I) \quad (5)$$

the concentration of ionophore for which P is being calculated, δ_I is the chemical shift of the Tl^+-ionophore complex, and δ_f is the shift of the free or uncomplexed Tl^+. The values of δ_I and δ_f are adjusted until the variance between the experimental chemical shifts and the chemical shifts calculated using the average stability constant ratio is minimized. Using the same type of fitting procedure, the absolute value of the stability constant of a Tl^+-ionophore complex in the absence of M^+ can be calculated using equation 6.

$$K^{Tl} = \frac{1 - P}{P\{[I]_{tot} - (1 - P) [Tl^+]_{tot}\}} \qquad (6)$$

The relative stability constants of the complexes of the alkali metal ions, silver ion, and the ammonium ion with four crown ethers in three different non-aqueous solvent systems are shown in table 2 (17).

TABLE 2

Stability Constants of Crown Ether-Cation Complexes Relative to that of Tl^+

Crown	Solvent	Li^+	Na^+	K^+	Rb^+	Cs^+	NH_4^+	Ag^+
15C5	DMSO	0.9	1.4	1.1	1.3	1.0	0.5	0.6
18C6	DMSO	<0.1	0.3	16.3	8.8	1.4	0.4	<0.1
	DMF		0.1	3.3	1.4	1.0		
	MeOH		0.7	25.4	3.5	0.7		
DB18C6	DMSO	0.1	5.4	24.6	5.3	1.5	0.5	0.4
	DMF		6.3	15.5	2.8	0.8		
	MeOH		13.9	44.4	5.3	1.2		
DCH18C6	DMSO	0.6	0.1	11.0	5.2	2.1	0.4	0.5

The effect of the solvent on the binding constants can, in principle, be quantitatively determined using equation 1. In practice, however, too many of the terms are currently unknown to allow quantitative treatments to be made. In spite of this difficulty, certain trends in the data can be observed. The clearest example is provided by comparing binding constants in methanol, which is a poorly solvating solvent, with those in the more strongly solvating solvents, DMSO or DMF. In methanol, 18C6 binds Na^+ and Cs^+ ions equally and Rb^+ only slightly more strongly. In DMF, the solvation of the ions is much more important and probably follows the charge/radius trend of solvation $Na^+>Rb^+>Cs^+$. Thus, in DMF, the binding constants to the ionophore would be expected to be more widely separated, with sodium having a smaller value than rubidium. In addition, the binding constants of Rb^+ and Cs^+ to the ionophore would be expected to be much closer in magnitude to that of K^+ because the potassium ion should be more strongly solvated. Both of these trends are observed. Similar reasoning should apply to cation binding by DB18C6. In methanol, the observed order of the binding constants to the ionophore is $Na^+ > Rb^+ > Cs^+$. In the better solvating solvents, the binding constant of Na^+ would be expected to become smaller relative to those of Rb^+ and Cs^+ because the cation is more strongly solvated. Correspondingly, the binding constant of Rb^+ would be expected to become smaller relative to that of Cs^+. The measured binding constants support this reasoning. The relative values of the constants for Na:Rb:Cs are 11.6: 4.4:1 in methanol; 3.6:3.5:1 in DMSO, and 7.9:3.5:1 in DMF.

IV. ORGANOTHALLIUM COMPOUNDS

The thallium chemical shifts and thallium-proton coupling constants in the linear dimethylthallium ion are solvent and anion dependent. The solvent dependence of the chemical shift is over 200 ppm and correlates linearly with the thallium-proton coupling constant. In order to determine the solvent properties which are important in the solvation of dimethyl thallium salts, an effort was made to correlate various solvent properties and parameters with the thallium chemical shifts. The only linear correlation was found between the shifts and the pK_a of the solvent. Parameters such as dipole moment, molar polarizability and dielectric constant do not correlate with the shifts (2).

The chemical shifts of many metal ions in solution have been interpreted in terms of ion pairing. A detailed model based on site symmetry and ion pairing was proposed to explain the solvent dependence of the chemical shifts observed for a variety of thallous salts (3). In order to determine the importance of ion pairing on the chemical shifts of the dimethylthallium salts, the difference in the shifts for the perchlorate and nitrate salts in a solvent were plotted versus the dielectric constant for that solvent. A linear correlation is observed as shown in fig. 2. The correlation coefficient for the line by linear regression analysis is R = 0.95.

The observed linear correlation between the shift difference and the dielectric constant of the solvent is indicative of the importance of ion pairing on the chemical shift. By plotting the difference in shift for two salts in a series of solvents, the effect of the solvation of the dimethylthallium ion is cancelled. As the dielectric constant increases, the ion pairing decreases and the difference between the chemical shift of the nitrate and perchlorate salts decreases. Because of the linear relation between the thallium proton coupling constant and the chemical shift, the plot also shows the importance of ion pairing on J_{Tl-H}.

The change in the chemical shift caused by phenyl group

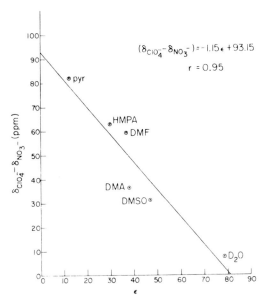

$$(\delta_{ClO_4^-} - \delta_{NO_3^-}) = -1.15\epsilon + 93.15$$

$$r = 0.95$$

Fig. 2. Thallium chemical shift difference vs. the solvent dielectric constant.

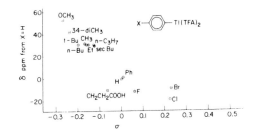

Fig. 3. Effect of substituents on the ^{205}Tl chemical shift plotted as the shift from arylthalliumbis(trifluoroacetate) vs. the Hammett σ constant.

substituents in arylthallium compounds are shown in fig. 3 plotted against the Hammett sigma constant. The chemical shift range is 55 ppm, roughly four times that observed using ^{13}C nmr. As expected, the chemical shift is extremely sensitive to subtle substituent effects and can distinguish between the various alkyl substituents such as methyl and ethyl. Similar studies have shown that ^{29}Si, ^{13}C, ^{19}F, ^{1}H, and ^{15}N exhibit increased shielding with increased electron donation of the substituents (18). In contrast, ^{31}P, (19) ^{199}Hg, (20) and now ^{205}Tl (21) have decreased shielding with increasing electron donation. In the case of ^{31}P, Gutowski and McCall postulated that double bond character influences the observed trend (22). On the other hand, in the case of the arylthallium compounds, changes in the dissociation of one of the ligands from the thallium or changes in the geometry around the thallium caused by the changing electron donating ability of the substituents could also influence the shift. Further studies are in progress to explain the observed correlation.

V. REFERENCES

(1) Dechter, J.J., and Zink, J.I., J. Amer. Chem. Soc., 97, 2937 (1975).

(2) Schramm, C., and Zink, J.I., J. Mag. Res., 26, 000 (1977).

(3) Dechter, J.J., and Zink, J.I., Inorg. Chem., 15, 1690 (1976).

(4) Erlich, R.H. Greenberg, M.S., and Popov, A.I., Spectrochim. Acta, 29, 543 (1973).

(5) Covington, A.K., Lilley, T.H., Newman, K.E., and Porthouse, G.A., J. Chem. Soc., Faraday Trans., 1, 963 (1973).

(6) Drago, R.S., Vogel, G.C., and Needham, T.E., J. Amer. Chem. Soc., 93, 6014 (1971).

(7) Gutmann, V., "Coordination Chemistry in Non-Aqueous Solutions", Chapter, Springer-Verlag, New York, 1968.

(8) Christensen, J.J., Eatough, D.J., and Izatt, R.M., Chem. Rev., 74 , 351 (1974).

(9) Chock, P.B., and Titus, E.O., Prog. Inorg. Chem., 18, 287 (1973).

(10) Pressman, B.C., Inorg. Biochem., 1, Chapter 6, 203 (1973).

(11) Truter, M.R., Ibid., 16, 71 (1973).

(12) Simon, W., Morf, W.E., and Meier, P.Ch., Ibid., 16, 113 (1973).

(13) Izatt, R.M., Eatough, D.J., and Christensen, J.J., Ibid., 16, 161 (1973).

(14) Lardy, H., Fed. Proc., Fed. Amer. Soc., Exp. Biol., 27, 1278 (1968); Eisenman, G., Ciani, S.M., and Szabo, G., Ibid., 27, 1289 (1968); Tosteson, D.C., Ibid., 27, 1269 (1968); Eisenman, Ibid., 27, 1249 (1968).

(15) Pressman, B.C., Ibid., 27, 1283 (1968).

(16) Dechter, J.J., and Zink, J.I., J. Amer. Chem. Soc., 98, 845 (1976).

(17) Srivanavit, C., Zink, J.I., and Dechter, J.J., J. Amer. Chem. Soc., 99, 0000 (1977).

(18) Scholl, R.L., Maciel, G.E., and Musher, W.K., J. Amer. Chem. Soc., 94, 6376 (1972).

(19) Mitch, C.C., Freedman, L.O., and Moreland, C.G., J. Mag. Res., 3, 446 (1970).

(20) Borzo, M., and Maciel, G.E., J. Mag. Res., 19, 279 (1975).

(21) Similar data appeared while this work was in progress.
      Hinton, S.F., and Briggs, R.W., J. Mag. Res., 22, 447
      (1976).
(22) Gutowski, H.S., and McCall, D.W., J. Chem. Phys., 22,
      162 (1954).

COMPLEXES OF PHOSPHORAMIDES AND POLYPHOSPHORAMIDES WITH CO(II) AND THIOCYANATE IN AQUEOUS SOLUTIONS

Yehuda Ozari* and Joseph Jagur-Grodzinski
Plastics Research Department
Weizmann Institute of Science
Rehovot, Israel

ABSTRACT

The formation of the Co(II)-thiocyanate-phosphoramide complexes in aqueous solutions was investigated. Polarographic experiments indicated stabilization of the complexed Co(II). The complexing powers of several phosphoramides and their polymeric analogues were determined using spectrophotometric methods. The complexing power of a polyphosphoramide calculated directly from components separation by ultrafiltration supports these results. Polydentate polymers are much more effective than the monomers and Co(II) affinity of the monophosphoramides increases in the order of their increasing Donor Numbers. The affinity of the polyphosphoramides increased with the polarity of their phosphoryl groups and with their molecular weight. Diffusion rates of Co(II) and thiocyanate in water insoluble phosphoramidic membranes were determined from the changes of their optical density. It is suggested that the complexes formed have tetrahedral symmetry; cobalt ion is first surrounded by four NCS anions, which form its first coordination shell, while phosphoryl groups form the second coordination sphere. According to kinetic considerations, it was noted that in such systems the phosphoramides and the thiocyanate act as synergists.

INTRODUCTION

Several investigators have pointed out that Co(II) forms tetrahedral complexes with hexamethyl phosphoramide (HMPA) in non-aqueous media (1-4). Co(II) complexes of similar structure with isothiocyanate anions in non-aqueous (5) and in aqueous solutions (6) have also been described. Measurable quantities of this complex in water could be observed only

* GAF Corporation, 1361 Alps Road, Wayne, N. J. 07470

in the presence of a large excess of thiocyanate since its
formation constant is extremely low (6-8). The absorption
band at 616 nm has been ascribed (6,7) to the tetrahedral [Co
(NCS)$_4$]$^=$ complex. It has been recently noted (9) that the
properties of the aqueous Co(II)-thiocyanate solutions are
highly affected by the presence of HMPA and other phosphoramides. While addition of 0.5 mole of NaSCN to Co(SO$_4$)$_2$ solution does not induce significant absorption in the 600-650 nm
region, an intense blue absorption band (at 627 nm) does
appear upon subsequent addition of 0.25 mole of N,N',N"-tris
(tetramethylene)phosphoramide (TPPA) (Fig. 1). Such solutions also have a broad absorption maxima at 1300 nm. Addition of anions other than thiocyanate has a negligible effect
on the spectra.

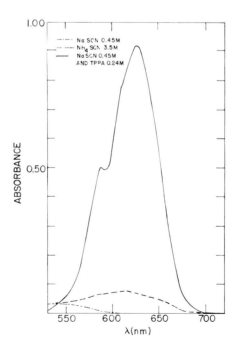

Fig. 1. *Optical spectra of aqueous*
CoSO$_4$ solutions (5.05 x 10^{-4} M) in
the presence of NH$_4$SCN or NaSCN and
TPPA.

The electronic spectra of the aqueous Co(II)-thiocyanate-phosphoramide complex is essentially identical with that of the tetrahedral Co(II)-isothiocyanate complex (5,6) in organic solvents. It is also similar, though more intense, to that of Co(II) complexes isolated from organic solvents with HMPA and NCS anions (1,3,4) or with HMPA molecules only as ligand species (12,13). This absorption corresponds to the electronic transitions calculated on the basis of ligand field strength theory for tetrahedral Co(II) complexes (12, 14).

The effect of HMPA on aqueous Co(II)-thiocyanate system was demonstrated also by polarographic measurements (9). The half wave reduction potential of the uncomplexed Co^{2+} (10^{-3}M) (-1.09 V vs. SCE) increases in the presence of thiocyanate anions (0.5 M) (-1.43 V vs. SCE) which is attributed to the reduction of $(CoSCN)^+$ (15). An additional increase occurs in the reduction potential in the presence of HMPA (0.2-0.7 M) (-1.65 V vs. SCE) which should be attributed to the reduction of $(CoSCN)^+$ $(HMPA)_n$ complex.

HMPA is a powerful donor (10,11); it is expected, therefore that it might replace water molecules in the coordination sphere of the cation. Phosphoramides of a donor strength higher than that of HMPA have been described recently (16) and the complexing power of polydentate ligands is much higher than that of their monofunctional analogue (17). Since the symmetry of these complexes seems to be well documented, the main aim of the present investigation was to determine the composition and the stability constants of these monomeric and polymeric phosphoramide complexes.

EXPERIMENTAL

Instruments

Hitachi Vapor Phase Osmometer and Ubelholde Viscosimeter were used in molecular weight determinations. Cary 14 Spectrometer was used in the near IR and Cary 15 in all other electronic absorption determinations.

Materials

HMPA (Fluka, practical) was left overnight over 4A Linde Molecular Sieves and distilled in vacuum (76°C at 10^{-3} torr.) N-tetramethylene-N',N"-bis(dimethyl)phosphoramide (MPPA), N,N'-bis(tetramethylene)-N"-dimethyl phosphoramide (DPPA, N,N'N"-tris(tetramethylene) phosphoramide (DPPA), N,N',N"-tris(tetramethylene) phosphoramide (TPPA) and poly-TPPA were synthesized and purified as described in ref. 18. Poly-HMPA was synthesized as described in ref. 19. Water soluble phosphoramides were obtained by grafting N,N'-bis(dimethyl)-N"-(diethylene amide) phosphoramide (TMAP) or N,N'-bis(tetra-

methylene)-N''-(diethylene amine) phosphoramide (DPAP) on
methyl-brominated polyphenylene oxide (9). Membranes were
cast from their solutions in 1,1,2-trichloroethane.

Ultrafiltration

Solutions of Co(II)-poly-TPPA-thiocyanate complexes were
filtrated through cellophane membranes (SGA Scientific Inc.
D-2130) under nitrogen pressure (400 psi).

Co(II) Analysis

1-2 ml of Co(II) solution ($10^{-2} - 10^{-3}$ M) was poured in
10 ml volumetric flask. 5 ml of 60% NH_4SCN in water, 1 ml
Na_2HPO_4 0.25 M (as a buffer) and 0.6 ml of TPPA were added.
The mixture was diluted to 10 ml and its absorption was
measured at 627 nm and referred to a calibration curve.

Diffusion

The membrane was attached to an inert clamp fitted for a
Cary 15 spectrometer. The membrane with the clamp was im-
mersed in the Co(II)-thiocyanate solutions for measured inter-
vals of time, washed with NH_4SCN solution (3.5 M) and the membrane absorption was measured at 627 nm.

RESULTS AND DISCUSSION

The formed Co(II) complex comprises both phosphoramide,
P, and thiocyanate ions.

$$Co^{2+} + nP + m^-SCN \rightleftharpoons \left[CoPn(NCS)_m\right]^{(m-2)-} \qquad (1)$$

Accordingly,

$$("C")/ \left[(Co^{2+}) (P)^n (^-SCN)^m \right] = Kov \qquad (2)$$

where, $("C")$, (P), (Co^{2+}) and (^-SCN) are activities of the
complex, of a phosphoramide, of free Co(II) and of thiocyanate ions, respectively.

It follows from Eq.2 that for a constant $(^-SCN) \gg [Co^{2+}]_o$

$$("C")/ \left[Co^{2+}) (P)^n\right] = b_n'' \qquad (2a)$$

and for a constant $\left[P_o\right] \gg \left[Co^{2+}\right]_o$

$$("C")/ \left[(Co^{2+}) (SCN)^m \right] = b_m' \qquad (2b)$$

We assume: 1) the optical absorption, I, at 627 nm is due to
the complex $"C"$ only and its maximum intensity, I_{max}, is determined by the value of $\left[Co^{2+}\right]_o$. This is justified for
the following reasons: a) the concentration or the type of
phosphoramide used do not affect the absorption band (627 nm);
b) at higher thiocyanate and phosphoramide concentrations,

its intensity reaches a plateau value $(E_c{}^{627}= 1880, E_c{}^{1330} = 680)$; c) the spectrum is virtually identical with that of $\left[Co(NCS)_4\right]^{-2}$ (5,6).   Since the absorbances of other thio-cyanate complexes of Co(II) at this wavelength are at least three orders of magnitude lower (6), their contribution to this absorption may be neglected. 2) "C" and Co^{2+} only must be considered in order to account for the total concentration of Co(II).

It follows from these two assumptions that: $("C")/Co^{2+})= I/(I_{max}-I)$.

As expected from Eq. 2a, straight and nearly parallel lines were obtained when values of $log\left[I/I_{max}-I)\right]$ were plot-ted against $log\ P$ for various phosphoramides at a high and constant concentration of thiocyanate (3.5 M). With the ex-ception of MPPA at $22^{\circ}C$ (slope $= 6$), the slopes of the lines shown in Fig. 2 are very close to 4. A slope 4 is also ob-tained for MPPA at $50^{\circ}C$.

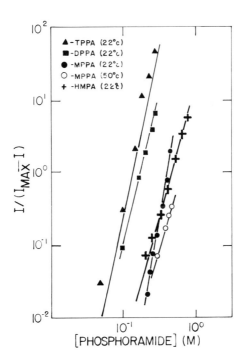

Fig. 2. $Log\ I/(I_{max}-I)$
$(I\ at\ 627\ nm)\ vs.\ log\ phosphoramide$
$concentration\ (NH_4SCN = 3.5\ M,$
$CoSO_4 = 10^{-3}M)$.

When P was kept constant (1.1 M HMPA or 0.5 M TPPA) the plot of $I/(I_{max}-I)$ vs. (^-SCN) gave an initial slope 4 $(b'_4 = 4100$ $M^{-4})$ (Fig. 3) and at concentrations of NaSCN higher than 0.1 M, the slope approaches a value of 2 $(b_2' = 41$ $M^{-2})$. Co(II) thiocyanate complexes of various stoichiometries were identified spectroscopically (6-8) and their presence was confirmed by nmr measurements (20). Their distribution was found to be dependent on the thiocyanate concentration. This explains the curvature noted in Fig. 3. The ratio $b_4'/b_2' \approx 100$ which corresponds to the global formation constant, b_2, for the reaction $Co^{2+} + 2^- SCN \rightleftharpoons Co(SCN)_2$, is in excellent agreement with the value of $b_2 = 100 \pm 20$ reported in the literature (21). Co(SCN)$_2$ prevails at the 3.5 M NH$_4$SCN, before addition of phosphoramide, and its conversion into the final tetrahedral complex represents the main reaction reflected by the results of those experiments.

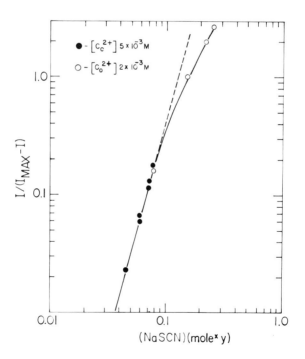

Fig. 3 Log I/I$_{max}$$^-$I)
(at 627 nm) vs. log NaSCN
activity. (HMPA = 1.1 M).

The overall stoichiometry of the presently investigated complexes *(n=4, m=4)* and the identity of their spectral characteristics with those of the tetrahedral $[Co(NCS)4]^=$ indicate that their first coordination shell is formed by the isothiocyanates only, while phosphoramides are restricted to the second coordination sphere. The small effect of complexation on the pmr spectra of the phosphoramides and the sensitivity of the phosphoramide coordination numbers to temperature changes support this conclusion. The phosphoramides replace the water molecules and stabilize the thiocyanate complex. This may be explained on kinetic grounds. It was noted (22) that the complexation of a transition metal ion enhances the exchange rate of the remaining solvent molecules. Thus, partial complexation of Co(II) with NCS anions may enhance the exchange rate of water by phosphoramide, which in turn will enhance the exchange of remaining water molecules by thiocyanate groups.

The mathematical treatment was based on the assumption that one coordination number of the phosphoramide mainly exists and that the contribution of other phosphoramide stoichiometries to the spectra might be neglected. Therefore, an effort was made to find out how extensive such contributions may be. A series of equilibria can be written:

$$A + P \overset{b_1}{\underset{}{\rightleftharpoons}} AP_1 \;\ldots\; A + nP \overset{b_n}{\underset{}{\rightleftharpoons}} AP_n$$

where P and b_n are concentrations of phosphoramide and global equilibrium constants respectively. A represents fractions of free Co^{2+} and $Co(II)$ complexes without phosphoramide which were fixed by using a large excess of NH_4SCN (3.5 M). E_O denotes the corresponding average molar extinction coefficient at 627 nm of these fractions.

Since for $P \gg [Co^{2+}]_o$ $\quad P_n = C_n / (AP^n)$

$$[Co(II)]_o = A + Ab_1P^1 + \;,\ldots,\; + Ab_nP^n \tag{3}$$

$$I = AE_O + AE_1b_1P^1 + \;,\ldots,\; + AE_nb_nP^n \tag{4}$$

$$I_o = [Co(II)]_oE_O \tag{5}$$

where E_i's are the molar extinction coefficients of the various complexes. It follows from Eqs. 3-5 that

$$I/[Co(II)] = B_i = (E_ob_1P_iE_1 + \;,\ldots,\; + b_nP^nE_n)/(1 + b_1P_i +$$

$$+,\ldots,\; + b_nP_i^n) \tag{6}$$

Terms of order higher than 6 were neglected. Eq. 6 contains 12 parameters: $b_1 \ldots b_6$ and $E_1 \ldots E_6$. In order to determine the values of these parameters 12 B_i values were obtained by plotting I vs. $\left[Co^{2+} \right]_o$ at 12 different constant P_i's of HMPA (0.2-0.8 M range). Then Eq. 6 was rewritten as a 12 parameter matrix and solved using a computer program based on least square iteration procedure. The slight deviation from zero (0.025) of the computed values $(F_i - B_i)/B_i$ (where F_i is the right hand side of Eq. 6) indicates the reliability of the solution. A perfect agreement exists between the computed E_4 (1872) (Table 1) and E (1880) derived from the plateau of the absorption curves. The dominant coordination number is four and b_4 coincides with b'' derived from the HMPA plot shown in Fig. 2. Comparison of K_n's values of the individual complexation steps (Table 1) suggests that the introduction of the first phosphoramide brings about an enhancement of the subsequent complexation step. Also, the more symmetric complexes comprising an even number of HMPA molecules are favored over those with an odd number.

Table 1

Equilibrium Constants, K_n, Global Equilibrium Constants, B_i, and Molar Extinction Coefficients of Co(II)-thiocyanate-HMPA Complexes formed in Aqueous Solutions in the Presence of 3.5M NH_4SCN. Temp. $22^{\circ}C$.

Number of attached HMPA Molecules	1	2	3	4	5	6
E_{627}	190	966	738	1872	2069	181
B_n	$.0009M^{-1}$	$.09M^{-2}$	$.3M^{-3}$	$20.9M^{-4}$	$.027M^{-5}$	$.0106M^{-6}$
$K_n (M^{-1})$	9.0×10^{-4}	100	3.3	69.7	1.3×10^{-3}	.4

In poly-HMPA and poly-TPPA the phosphoryl groups are built into their backbone and spaced at about 6 Å. Because of a possible cooperative effect of the neighboring groups, it is expected that the polymers might act as much more effective complexing agents than the monomers. The *log-log* plot in Fig. 4 yields straight lines of 45° slopes for both polymer systems. The identity of the spectral characteristics of the polymeric complexes with those of the monomeric ones seems to indicate that interpretation of 1:1 stoichiometry (based on 45° slopes) is incorrect.

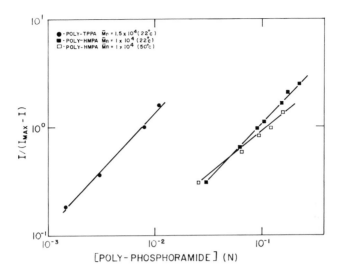

Fig. 4. Log $I/(I_{max}-I)$
(I at 627 nm) vs. log normalty
of polyphosphoramide (in respect
to phosphoryl groups). (NH_4SCN =
3.5 M, $CoSO_4 = 10^{-3}$ M).

Let us consider a system in which the overall complexation equilibrium is determined by the rates of reactions involving the attachment of the first complexing entity and the removal of the last one from the central atom. Obviously, 1:1 apparent stoichiometry will be indicated by the concentration dependence of the equilibrium. It is reasonable to postulate that such behavior characterizes the investigated polymeric systems. After the polydentate molecule is attached by one phosphoryl group to the Co(II), the probability of having additional phosphoryl groups in its vicinity is not determined by their overall concentration, but by their position within the polymeric chain. Thus, the equilibria involving additional complexation steps become concentration-independent. Hence, an apparent 1:1 stoichiometry of the complex.

The reliability of the results obtained with poly-HMPA was further checked, as follows. For a 1:1 stoichiometry and for a high and constant concentration of thiocyanate ions, Eq. 2 can be written in the form:

$$1/P = -b'' + (b''E\left[Co^{2+}\right]_o)(1/I) \qquad (8)$$

Thus, both b and E can be simultaneously derived from a series of experiments at a constant $\left[Co^{2+}\right]_o$. Linear plots of $1/P$ vs. $1/I$ obtained yield $b'' = 10$ M^{-1} and 80 M^{-1} and E values of 1775 and 1760 corresponding to $\overline{M}n = 10^4$ and 2.5×10^4 respectively. The E's are in excellent agreement with that of the "C" complex derived from the plateau values of the experiments conducted with other phosphoramides.

The results obtained from the spectrophotometric methods were examined for a poly-TPPA system by component separation by ultrafiltration. Poly-TPPA was dialyzed through cellophane films such that the obtained polymer did not pass through the pores of the membrane. In the ultrafiltration process of the aqueous solution of the "C" complex ($^-$SCN = 3.5 M) the uncomplexed Co(II) should pass through the membrane, while the complexed Co(II) remains in the solution, attached to the polymer. If the volume of the filtrate is small enough such that the equilibrium is not changed during the process, its Co(II) concentration should be, basically, equal to the uncomplexed Co(II). A correction was made due to Co(II) rejection (R_O) by the membrane in the absence of polymer $(R_O \approx 10\%)$.

$$R_o = 1 - (Co\ filtrate)/(Co\ total) \qquad (7)$$

$$(Co^{2+}) = (Co\ filtrate)/(1 - R_o) \qquad (8)$$

Values of (Co^{2+}) obtained for different concentrations of poly-TPPA $(4.18$-$19.60) \times 10^{-3}$ M were substituted in Eq. 2**a** $\left[\text{for } n = 1 \text{ and "C"} = (C_{total}) - (Co^{2+})\right]$ and yielded b'' 137.5 $-$ 148.3 M^{-1} which agree very well with that of the spectrophotometric experiments (149 M^{-1}) (Table 2).

The equilibrium constants calculated for monomeric and polymeric species are summarized in Table 2. The Donor Numbers of the investigated phosphoramides increase in the order (16): HMPA $\leq$ MPPA < DPPA < TPPA. As can be seen from the tabulated data, their respective abilities to complex Co(II) increase in the same order.

Table 2

Global Equilibrium Constants of Complexation with Co(II) of Various Phosphoramides in Aqueous Solutions of 3.5M NH_4SCN.

No	Phosphoramide	Coordination Number		b''	
		at 22°C	at 50°C	at 22°C	at 50°C
1	HMPA	4	–	$21.4M^{-4}$	–
2	MPPA	6	4	$1680.0M^{-6}$	$166M^{-4}$
3	DPPA	4	3	$1200M^{-4}$	$204M^{-3}$
4	TPPA	4	–	$4000M^{-4}$	–
5	Poly-HMPA $\overline{M}n$-10^4	1	0.8	$10M^{-1}$	$8.4M^{-0.8}$
6	$\overline{M}n$=2.5×10^4	1	–	$80M^{-1}$	–
7	Poly-TPPA $\overline{M}n$=1.5×10^4	1	–	$149M^{-1}$	–

A comparison between the complexing power of the monomeric and polymeric species based directly on the global equilibrium constants would be misleading, because they refer to different apparent stoichiometries. It seems, however, that
values of b'' for the polyphosphoramides should be compared
with values of K_1 for HMPA (Table 1). Such comparison reveals that the investigated systems are not different in this
respect from other polydentate complexing agents (17), which
were shown to act as much more effective than their monoden-
tate analogues.

The effectiveness of the polyphosphoramides seems to be
also dependent on their molecular weight. An eight-fold increase in b'' was noted when the molecular weight of poly-HMPA
was increased from 10,000 to 25,000. Similar to the monomer
series, poly-TPPA is again a more powerful complexing agent
than poly-HMPA of the same molecular weight. It seems reasonable to assume that with further increase in molecular weight
the complexing power would eventually reach a plateau value.

The apparent enthalpies and entropies of the complexation
reactions were derived from the semilog plots of b'' vs. recip-
rocals of absolute temperatures (Fig. 5). ΔH values were
-3.46 Kcal/mole for TPPA and -2.85 Kcal/mole for poly-HMPA
independent of its molecular weight. The larger value of ΔH
for TPPA is in agreement with the high basicity of its phosphoryl groups. For TPPA, the entropy of complex formation is
positive ($\Delta S = 5.1$ e.u.). Such value indicates that in this
system the formation of the complex is accompanied by an increase of disorder in the bulk of the solution, probably dis-
ruption of TPPA clusters. On the other hand, for complexes
involving poly-HMPA molecules, the entropy values were nega-
tive (-1.55 e.u. and -4.44 e.u. for $\overline{M}_n=2.5\times10^4$ and 10^4, respectively). The increase in complexing ability with the increase
in the molecular weight is related to the decrease in the
entropy losses due to complexation. Apparently, the phosphoryl groups in the longer polymer are organized around the
central atom in such a way that the resulting loss of segment-
al mobility is significantly reduced.

Polarographic experiments were conducted at a constant
NaSCN concentration while the concentration of HMPA was
changed. Above 0.2 M of HMPA, the half reduction potential of
Co(II) becomes a constant (-1.65V vs. SCE). This value should
be compared to -1.09 and -1.43V vs. SCE for Co^{+2} and cobalt
thiocyanate complex, respectively. According to ref. 15, this
reduction wave should be attributed to the reduction of the
$(CoSCN)^+$ at the mercury interface and in our case, it is the
$(CoSCN)^+$ $(HMPA)_4$ complex. The spectroscopic and the polarographic results should be referred to different species, therefore numberical values derived from the two methods could not
be compared. Nevertheless, these experiments demonstrate that
significant changes occur in the cobalt thiocyanate solution
as a result of addition of small amounts of phosphoramides.

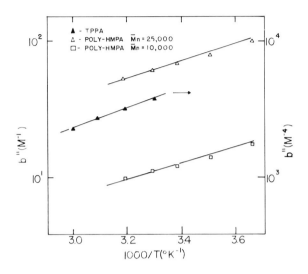

Fig. 5. Semi-log plots of the formation constants vs. reciprocal of the absolute temperature.

Two types of water insoluble phosphoramide polymers were synthesized: dimethylamine and pyrrolidine substituted phosphoramides (TMAP and DPAP, respectively). Membranes of these polymers absorb selectively Co(II) from its diluted aqueous solutions ($10^{-3} - 10^{-5}$ M) containing thiocyanate ions. Their Co(II) exchange capacity could be controlled by varying the percentage of the phosphoryl groups. Membranes with capacity of 0.5 meg/g were used. This value corresponds to 4:1 phosphoramide — Co(II) stoichiometry, which is identical with the spectrophotometric results for water soluble phosphoramides. The spectral characteristics of the "C" complex in the membrane are identical to those in the aqueous solution. Therefore, the absorption of Co(II) could be determined directly from changes of the optical density of the membranes at 627 nm. The diffusion coefficients of Co(II) in DPAP membranes ($D = 0.2 \times 10^{-10}$ cm^2 sec^{-1}) are ~20 times lower than those in TMAP membranes ($D - 4.0-6.0 \times 10^{-10}$ cm^2 sec^{-1}). It appears that this difference can be attributed to differences in the respective rates of decomposition. Indeed, the stability constant of the DPAP seems to be higher than the TMAP. The effectiveness of the membranes in Co(II) absorption is demonstrated by the relatively low concentration of the thiocyanate (10^{-1} M) in the solution needed to attain maximum Co(II) capacity (Fig. 6). The membrane containing more polar phosphoryl

groups (DPAP) is more effective in Cobalt(II) absorption in
low concentration region of thiocyanate but the plateau value
attained for both polymers (TMAP and DPAP) is the same . The
effectiveness of the membranes in Co(II) absorption is explained by the low dielectric constant of the membrane media
which associates Co(II) and thiocyanate ions, which enhances
the "C" complex formation.

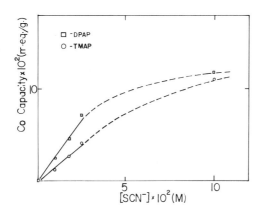

*Fig. 6. Co(II) capacity of water
insoluble resins vs. thiocyanate
concentration. (CoSO$_4$ = 10^{-2} M).*

References

1. Donaghue, J.T. and Drago,R.S., Inorg. Chem. 2, 572 (1963).
2. Sohafer, M. and Curran, C., ibid., 4, 623 (1965).
3. Le Coz, E., Guerchais, J.E. and Goodgame, D.M.L., Bull.
    Soc. Chim. France 3855 (1969).
4. (a) De Bolster, M.W.G. and Groeneveld, W.L. Recueil Tral.
    Chim. Pays-Bas 90, 477 (1971).
    (b) Idem., 91, 171 (1972); (c) idem., 91, 643 (1972).
5. Cotton, F. A., Goodgame, D.M.L., Goodgame, M., and Sacco,
    A., J. Am. Chem. Soc. 83, 4157 (1961).
6. Tribalat, S and Zeller, C., Bull. Soc. Chim. France 2041
    (1962).
7. Lehne, M., ibid., 76 (1951).
8. Babko, A. K. and Drako, O.F., Zh. Obsh. Khim. 20, 228
    (1950).
9. Ozari, Y., Ph.D. Thesis, Weizmann Institute of Science
    (1975).
10. (a) Normant, H., Angew. Chem. Internat. Edit. 6, 1046
    (1967).
    (b) Normant, H., Russian Chem. Rev. 39, 357 (1970).

11. Gutmann, V. and Scherhaufer, A., _Monatsh_. 99, 335 (1968).
12. Donaghue, J.T. and Drago, R.S., _Inorg. Chem_. 1, 866 (1962).
13. Gutmann, V., Weisz, A., and Kerber, W., _Monatsh_. 100, 2096 (1969).
14. Cotton, F. A. and Goodgame, M., _J. Am. Chem_. Soc. 83, 1777 (1961).
15. Perone, S. P. and Gutknecht, W. F., _Anal. Chem_. 39, 892 (1967).
16. Ozari, Y. and Jagur-Grodzinski, J., _J.C.S. Chem. Comm_. 295 (1974).
17. Chan, L.L. and Smid, J., _J. Am. Chem. Soc_. 89, 4547 (1967).
18. Ozari, Y. and Jagur-Grodzinski, J., submitted to _Inorg. Chem_. (1976).
19. (a) Bello, A., Bracke, W., Jagur-Grodzinski, J. Sackmann, G and Szwarc, M., _Macromolecules_ 3, 98 (1970). (b) Ozari, Y. and Jagur-Grodzinski, J., _J.C.S_. Dalton, 474 (1973).
20. Horrocks, W.D.Jr. and Hutchison, J.R., _J. Chem. Phys_. 46, 1703 (1967).
21. Tribalat, S and Caldero, J. M., _J. Electroanal. Chem_. 5, 176 (1963).
21. Plotkin, K., Copes, J.and Vriesenga, J. R., _Inorg. Chem_. 12, 1494 (1973).

AUTHOR INDEX

SUBJECT INDEX

A
B
C 8
D 9
E 0
F 1
G 2
H 3
I 4
J 5